Innovation as Social Change in South Asia

This book examines innovation as social change in South Asia. From an anthropological micro-perspective, innovation is moulded by social systems of value and hierarchy, while simultaneously having the potential to transform them. Peterson examines the printing press's changing technology and its intersections with communal and language ideologies in India. Tenhunen explores mobile telephony, gender, and kinship in West Bengal. Uddin looks at microcredit and its relationship with social capital in Bangladesh. Jeffery surveys imbalanced sex ratios and the future of marriage payments in north-western India. Ashrafun and Säävälä investigate alternative dispute resolution as a social innovation which affects the life options of battered young wives in Sylhet, Bangladesh.

These case studies give insights into how the deeply engrained cultural models and values affect the forms that an innovative process can take. In the case of some South Asian societies, starkly hierarchical and holistic structures mean that innovations can have unpredictable sociocultural repercussions. The book argues that successful innovation requires taking into account how social hierarchies may steer their impact.

This book was originally published as a special issue of *Contemporary South Asia*.

Minna Säävälä is an Adjunct Professor of Social Anthropology in the University of Helsinki, Finland, and works as a senior researcher in the Population Research Institute, Finland. Her current research projects relate to family formation in India and reproductive health of migrant populations in Europe.

Sirpa Tenhunen is an anthropologist who teaches in the University of Jyväskylä, Finland, as a Professor (interim) and University Lecturer. She has also taught anthropology in the University of Helsinki, Finland, and worked as a researcher in the Academy of Finland. In addition to new media, her research interests cover gender, work, and politics in India.

Innovation as Social Change in South Asia

Transforming hierarchies

Edited by
Minna Säävälä and Sirpa Tenhunen

Routledge
Taylor & Francis Group

LONDON AND NEW YORK

First published 2015 by Routledge

2 Park Square, Milton Park, Abingdon, Oxon OX14 4RN
711 Third Avenue, New York, NY 10017, USA

Routledge is an imprint of the Taylor & Francis Group, an informa business

First issued in paperback 2017

British Library Cataloguing in Publication Data
A catalogue record for this book is available from the British Library

ISBN 13: 978-1-138-85558-8 (hbk)
ISBN 13: 978-1-138-05976-4 (pbk)

Typeset in Times New Roman
by RefineCatch Limited, Bungay, Suffolk

Publisher's Note
The publisher accepts responsibility for any inconsistencies that may have
arisen during the conversion of this book from journal articles to book chapters,
namely the possible inclusion of journal terminology.

Disclaimer
Every effort has been made to contact copyright holders for their permission to
reprint material in this book. The publishers would be grateful to hear from any
copyright holder who is not here acknowledged and will undertake to rectify
any errors or omissions in future editions of this book.

Contents

Citation Information

The chapters in this book were originally published in *Contemporary South Asia*, volume 22, issue 2 (June 2014). When citing this material, please use the original page numbering for each article, as follows:

Chapter 1: Introduction
Innovation: transforming hierarchies in South Asia
Minna Säävälä and Sirpa Tenhunen
Contemporary South Asia, volume 22, issue 2 (June 2014) pp. 121–129

Chapter 2
Katibs and computers: innovation and ideology in the Urdu newspaper revival
Mark Allen Peterson
Contemporary South Asia, volume 22, issue 2 (June 2014) pp. 130–142

Chapter 3
Microcredit and building social capital in rural Bangladesh – drawing the uneasy link
Mohammad Jasim Uddin
Contemporary South Asia, volume 22, issue 2 (June 2014) pp. 143–156

Chapter 4
Mobile telephony, mediation, and gender in rural India
Sirpa Tenhunen
Contemporary South Asia, volume 22, issue 2 (June 2014) pp. 157–170

Chapter 5
Supply-and-demand demographics: dowry, daughter aversion and marriage markets in contemporary north India
Patricia Jeffery
Contemporary South Asia, volume 22, issue 2 (June 2014) pp. 171–188

Chapter 6
Domestic violence made public: a case study of the use of alternative dispute resolution among underprivileged women in Bangladesh
Laila Ashrafun and Minna Säävälä
Contemporary South Asia, volume 22, issue 2 (June 2014) pp. 189–202

Please direct any queries you may have about the citations to
clsuk.permissions@cengage.com

Notes on Contributors

Laila Ashrafun has been serving as a teacher of Shahjalal University of Science and Technology, Sylhet, Bangladesh, for the last 14 years. Her doctoral thesis *Seeking a Way Out of the Cage: Underprivileged Women and Domestic Violence in Bangladesh*, was published in 2013. Her current areas of interest are gender, family studies, HIV/AIDS and prostitution, and poverty.

Patricia Jeffery has been Professor of Sociology at the University of Edinburgh, UK, since 1996. Her research in rural north India since 1982 has focused on gender politics, childbearing, social demography, education, and communal politics. In 2009–2010 she was awarded a British Academy/Leverhulme Trust Senior Research Fellowship and Leverhulme Research Fellowship for work on a book about social, economic, and demographic change in western Uttar Pradesh. She is currently co-investigator in the ESRC-funded project on Rural change and anthropological knowledge in post-colonial India: A comparative 'restudy' of F. G. Bailey, Adrian C. Mayer, and David F. Pocock.

Mark Allen Peterson is Chair of Anthropology and Professor of International Studies at Miami University, Oxford, Ohio, USA. He is the author of the books *Anthropology and Mass Communication: Myth and Media in the New Millennium* (2003) and *Connected in Cairo: Growing Up Cosmopolitan in the Modern Middle East* (2011), and is co-author of *International Studies: An Interdisciplinary Approach to Global Issues* (2010). He is co-editor of the Anthropology of Media book series.

Minna Säävälä has been studying Indian society for the last 20 years. She is an Adjunct Professor of Social Anthropology in the University of Helsinki, Finland, and works as a senior researcher in the Population Research Institute, Finland. She is the author of *Fertility and Familial Power Relations* (2001), *Middle-class Moralities* (2010) and a co-author of *An Introduction to Changing India* (2012) with Sirpa Tenhunen. Her current research projects relate to family formation in India and the reproductive health of migrant populations in Europe.

Sirpa Tenhunen is an anthropologist who teaches in the University of Jyväskylä, Finland, as a Professor (interim) and University Lecturer. She has also taught anthropology in the University of Helsinki, Finland, and worked as a researcher in the Academy of Finland. In addition to new media, her research interests cover gender, work, and politics in India. Her recent books include *An Introduction to Changing India* (2012, with Säävälä), *Culture, Power and Agency: Gender in Indian Ethnography* (2006, with Fruzzetti), and *Means of Awakening: Gender, Politics and Practice in Rural India* (2009).

Mohammad Jasim Uddin has been serving as a public university teacher in Bangladesh for the last 15 years. His doctoral dissertation, *Micro-credit, Gender, and Neoliberal Development in Bangladesh*, was published in 2013.

INTRODUCTION

Innovation: transforming hierarchies in South Asia

Minna Säävälä[a,b] and Sirpa Tenhunen[c,d]

[a]Population Research Institute, Väestöliitto, Helsinki, Finland; [b]Department of Social Research, University of Helsinki, Helsinki, Finland; [c]Nordic Centre in India, Communication and International Relations Office, Umeå University, University Liaison Building, Umeå, Sweden; [d]Department of History and Ethnology, University of Jyväskylä, Jyväskylä, Finland

This special issue examines innovation as social change in South Asia. From an anthropological micro perspective, innovation is moulded by social systems of value and hierarchy and simultaneously potentially transforms them. The articles in this special issue examine a number of innovations in South Asian contexts: the printing press's changing technology and its intersections with communal and language ideologies in India (Peterson); mobile telephony, gender, and kinship in West Bengal (Tenhunen); microcredit and its relationship with social capital in Bangladesh (Uddin); imbalanced sex ratios and the future of marriage payments in north-western India (Jeffery); and how alternative dispute resolution as a social innovation affects battered young wives' life situation options in Sylhet, Bangladesh (Ashrafun and Säävälä). These case studies give insights into how the deeply engrained cultural models and values affect the forms that an innovation process can take. In a social field, actors are not situated symmetrically vis-à-vis an innovation. In South Asian societies that are starkly hierarchical and holistic, innovations may have unpredictable sociocultural repercussions.

Innovation, social and technological, is the current buzzword – also in South Asia. High hopes are harboured for both types of innovation as an engine of economic growth in the area. By boosting 'creativity, flexibility and adaptation', innovation could ultimately make, for example, India 'a global innovation leader' (Chidambaram 2007; S. Datta 2011). The ability to innovate has become a key factor in determining the competitive advantage of national economies. In South Asia – as elsewhere – the main interest of innovation studies has either been the dynamics of innovation within industry, government, and academia,[1] or innovation as an adoption process (Dearing 2009) or diffusion (Rogers 2003). However, innovations not only determine the competitiveness of national economies in the global market, but they also influence, and are influenced by, their particular local socioeconomic and historical contexts.

In this special issue, we examine innovations in South Asia from the vantage point of social change rather than as technological inputs for development and economic growth.

We regard innovation as a potential source of solutions for social problems in this subcontinent plagued by social asymmetries and hierarchies, whether due to poverty, gender, caste, or religion. Innovations are therefore too important socially to be only studied from technological, industrial, and governmental perspectives. The articles in this issue examine a number of innovations in South Asian contexts: the printing press's changing technology and its intersections with communal and language ideologies (Peterson); mobile telephony, gender, and kinship (Tenhunen); microcredit and its relationship with social capital (Uddin); imbalanced sex ratios and the future of marriage payments (Jeffery); and how alternative dispute resolution (ADR) as a social innovation affects battered young wives' life situation options (Ashrafun and Säävälä).

Innovation in whose interest?

Innovations, inventions, and change characterize all sociocultural wholes. In a sense, change is coded into the very existence and survival of a social entity. Change is essential but, simultaneously, a threat. In sociocultural systems, there is an ultimate need to find a balance between invention and convention. In the long run, cultural entities can only survive by immunizing themselves against radical change, but they simultaneously risk deterioration unless they manage to transcend what they are and transform themselves (Welz 2003). In some situations, change is a result of using traditional means of action and structures of meaning in novel and unpredictable situations. However, sometimes change takes place through a deliberate, intentional action – such as adopting an innovation. In the context of innovations, we mainly refer to intentional change, although the difference between the two modes of change is necessarily blurred. The case studies in this special issue show how the deeply engrained cultural models and values affect the forms an innovation process can take.

In a nutshell, an innovation can be defined as a new way of doing things that works. According to the classical definition, an innovation is an idea, practice, or object that the unit of adoption perceives as new (Rogers 2003). An innovation is, however, not equal to invention; an invention has to be successfully put into practice in order to become an innovation. The notion of innovation entails that it is useful for an actor's (or a 'unit of adoption's') interests. Innovation is only an innovation if it is successful – if it makes the situation better or is useful. This definition opens the issue to the question of power and legitimacy: for whom is the innovation useful or beneficial? Who determines whether the innovation works or is useful, whether it is an innovation in the proper sense of the term? In a social field, actors are not situated symmetrically vis-à-vis an innovation. As De Sardan (2005) argues, the introduction of an innovation is very likely to serve the interests of some, while damaging the interests of others.

Like Shove, Pantzar, and Watson (2012), who have analysed how innovations emerge, exist, and die, we are interested in the appropriation of innovations as social practices. Shove, Pantzar, and Watson (2012) define practices as consisting of materials, competences, and meanings; they argue that practices emerge, persist, shift, and disappear when connections between these types of elements are made, sustained, or broken. In contrast to their discussion on the social life of leisure practices such as Nordic walking and snowboarding, our focus on South Asian practices underlines the issues of inequality and local understandings of hierarchies. South Asian societies are plagued by a number of hierarchies – to such a degree that hierarchy was long considered as a defining feature of South Asian sociability (Dumont 1970). Although this emphasis on hierarchy and its ubiquitousness may have blinded academics to some South Asian realities (see critiques, e.g.

Appadurai 1986; Srinivas 1989; Chatterjee 1993; Gupta 2000), it is hard to think of social life in South Asia without the notion of hierarchy and asymmetry.

We explore how innovations relate to local hierarchies, whether these are kinship, class, religious, or gender hierarchies. Much of anthropological research on the uses of new technologies have an emphasis in common on how the appropriation of these technological innovations tends to reinforce existing structures (Horst and Miller 2006; Barendregt 2008; Archambault 2010). We ask whether innovations are mainly used to strengthen local hierarchies, or whether they also have a transformative potential. Our focus is more on the appropriation of innovations, although we also consider how they are constructed and made available in relation to social hierarchies, and how the appropriation of innovation relates to other ongoing social change processes. The majority of the articles (Tenhunen, Uddin, Jeffery, Ashrafun and Säävälä) touch upon gender and kinship hierarchies, while one (Peterson) refers to the hierarchical position of a religious minority, and another (Uddin) refers to hierarchies based on social capital and poverty.

The social life of innovation

From a social science perspective, all innovations are social. However, 'social innovation' is a particular concept that is important for understanding the role of innovations in development. In the world of liberalized and globalized economies, national governmental policies govern social development to a lesser degree. New forms of improving social equality and development, such as social innovations and third-sector initiatives, have become central in global international and local agencies' thinking (van der Ven et al. 1999; Bellemare and Boucher 2006; Pol and Ville 2009; Brown 2007; Wire 2007). Social innovation refers to any innovation that has a social purpose, for example, microcredit and social marketing strategies. Social innovations can be described as cross-sector collaborative endeavours that do not only 'belong' to the public sector, non-profit sector, or business sector, but are by definition inter-sectorial.

Social innovations – new ideas that further social ends and not economic or technical ends (Mulgan 2007) – specifically deal with the very core of the idea of development. If we, like Nussbaum (2000) and Sen (1999), understand development as a question of capabilities and entitlements, it is evident that a successful innovation should lead to social development. Whether this actually happens is a question of empirical analysis.

Technological innovations, too, can be transformative, affecting a wide range of social relations, politics, community, and kinship in a matrix that intersects these social fields with markets and ethical issues (Winslow 2009). Fundamentally, new inventions that depart radically from what went on earlier have to have 'interactive support systems' (Arthur 2005) in order to change mere inventions into innovations, into 'something that works.' Social institutions and cultural values are central for understanding how an innovation works or does not work and what kind of effects it has within its context (Winslow 2009). As social constructionist views of technology have illustrated, technologies do not merely consist of physical artefacts, devices, and infrastructures, but also of users' preferences, businesses, and regulatory policies (Pinch and Bijker 1984; Bijker and Law 1992; Bijker, Hughes, and Pinch 1993; Williams and Edge 1996). This special issue contains case studies on technological (Peterson and Tenhunen) and social innovations (Uddin, Jeffery, Ashrafun and Säävälä). By examining different kinds of innovations in the same regional context, we seek to provide a rich and nuanced insight into developmental processes and social change in South Asia.

South Asian innovative particularities

In South Asia, innovativeness and dynamic social forces particularly emerged after the liberalization of the economies: since the 1970s in Sri Lanka; the 1980s in Nepal, Pakistan, and Bangladesh; and since the beginning of the 1990s in India.[2] In this region, innovations are flourishing in the technology field, as well as in the social and organizational fields. Globally significant innovations have emerged here since the 1990s, the most prominent and influential of which was microcredit banking, which the Nobel Peace laureate Mohammed Yunus developed with the help of his Grameen Bank (Yunus 2007). Other examples of innovations include organizational innovations in health services and policies (Basu and Amin 2000; Richman et al. 2008; Dror et al. 2009; Padmanaban, Raman, and Mavalankar 2009), organizational innovations regarding urban contexts (Kundu 2001; Patel and Bartlett 2009), social innovations in business services (P. B. Datta 2011), and technological innovations in agriculture (Chhetri et al. 2012) and in pottery production (Winslow 2009). Numerous studies have analysed the emergence and use of innovations such as microfinance in South Asia (e.g. Bhatt 1995; Rahman 2001; Baruah 2002; Parmar 2003; Yunus 2007; Karim 2008), mobile telephony (e.g. Tenhunen 2008; Doron 2012; Tacchi, Kitner, and Crawford 2012; Jouhki 2013) and, for example, new technologies and practices in the field of human reproduction (e.g. Featherstone et al. 2005 on new genetics; Säävälä 1999 on female sterilization; and Pande 2009 on commercial surrogate motherhood). These studies are often not framed as studies of innovation, but as studies of new social situations that have emerged as an outcome of an innovation.

One of the most insightful micro studies of technological innovation in South Asia was Winslow's (2009) analysis of the long-term processes' effects on potters in her fieldwork area in Sri Lanka, which she followed since the 1970s. She witnessed the long process of producing pottery manually and with a potter's wheel to mechanization and the use of new technologies and marketing practices. In her analysis, she shows how the people in the village of Walangama deployed old ideologies of kinship and memory to maintain egalitarian and communitarian local, caste-based values and practices, while profitably engaging with Sri Lanka's post-structural adjustment economy. The Walangama potters' case study is a timely reminder that despite capitalism penetrating local communities, how a community reacts to the situation is not self-evident. Innovations and their social effects are not a simplistic story of ubiquitous commercialization and instrumentalization of social relations (Winslow 2009).

In this special issue, we provide the reader with an array of micro studies on innovation in South Asia. As Wajcman (2002) argues, technologies and innovations do not offer simple technological fixes for social problems, but are part of social changes in the way technologies are produced and socially used. We believe that by paying attention to details and down-to-earth social relations, it is possible to open new perspectives to understanding innovation in its socio-cultural context. Since technologies and innovations are culturally appropriated, they are most perceptible through an ethnographic lens (Winslow 2009). In turn, detailed attention to their multiple uses and influences can help create development interventions and policies which take the multiplicity of actors and ongoing social processes into account.

The contents of the special issue

Since India's market economy was revised to be more in line with global neoliberal practices in the early 1990s, stories of entrepreneurial success have been ubiquitous including that of the media's success. In these tales, media success – measured by growing audiences, ad revenues, and profits – is a direct result of the changing economy due to investment, credit capital, deregulation, and increasing technological efficiency. Mark Peterson seeks to challenge the simple technological and economic determinist explanations for social and cultural change; he does not contradict them, but emphasizes the significance of local cultural contexts. He argues that the current boom in Urdu newspapers cannot be entirely explained by economic liberalization, or print reproduction's new technologies. The revitalization of the Urdu press is a result of the intersection of these processes with shifting language ideologies and new writing technologies in a context of the Urdu-speaking peoples' changing attitudes to their communities. The innovations in India's new economy should thus not be seen as generally driven by market forces, deregulation, and new technologies, but also as encompassing historical trends specific to Urdu. These historical changes in Urdu language ideology have been decades in the making. The new understanding of the role of Urdu has been influenced by the changing meanings of a Muslim identity in India as well as by the contingencies of national political concerns.

Mohammad Jasim Uddin studies one of the most well-known and widely discussed innovations meant to reduce poverty and provide social capital: the group-based microcredit programmes that Dr Yunus and the Grameen Bank he founded in Bangladesh initiated. On the basis of case studies and ethnographic interviews with 151 female customers of the Grameen Bank and BRAC, 2 important microfinance institutions in Bangladesh, Uddin argues that the assumed association between microcredit membership and the building of social capital (social networks, norms of reciprocity, and collective identity and action) is much less important than many previous scholars and development practitioners commonly suggested. Uddin's article shows that microcredit organizations have gradually dropped those parts of their practices which could have had a transformative potential in order to ensure their money-lending business's growth and sustainability. Microcredit organizations no longer meticulously follow their agenda to provide microcredit to the poor and to ensure that female borrowers interact with one another and attend meetings. Since these women have not been given opportunities to invest their loans themselves, they commonly apply for microloans for their male family members' use.

Sirpa Tenhunen focuses on the transformative potential of mobile telephony in rural West Bengal, India. She illustrates how the physical qualities of phones help strengthen the multiplicity of discourses by mediating relationships and contributing to the multiplicity of speech contexts. She describes how mobile phone use has been encouraged and motivated by kinship relationships, and the use of mobile phones has, in turn, transformed these relationships by creating new contexts for speech and action. First, they enable translocal communication, helping callers to transgress social boundaries. Second, phones give callers new possibilities to choose the context for their speech and engage in critical and unconventional discourses, which can help women make tangible changes in their everyday lives. By enabling new contexts for speech, phones create possibilities to voice critical ideas which can challenge the power structure in a household. However, instead of drastic improvements or changes, for instance in economic power relationships, the positive impacts of women's phone use appear subtle and ambiguous: most calls are about the slight redefinition of the home boundaries.

Patricia Jeffery's essay discusses the potential repercussions of a process partly initiated by technical innovations: amniocentesis and ultrasound, which allow the sex of an unborn child to be determined. In the context of daughter aversion, this innovation has led to a process widely considered deleterious for women in India and as having deep societal and demographic effects. Jeffery takes a critical look at how the imbalanced child sex ratios, which the practice of foetus selection has aggravated, could affect marriage practices, particularly in the northern and north-western parts of India. Some demographers, economists, and social scientists have suggested that sex-selective abortions could lead to the growing value of girls as a 'scarce commodity' and a subsequent decrease in, or even annihilation of, the dowry practice in the subcontinent. Jeffery offers evidence that the homogenization, or disappearance, of the dowry practice is improbable; on the contrary, various marriage practices may in the future coexist in India, which will allow the privileged echelons of society to maintain the dowry practice, while the males of the lower socioeconomic groups may face difficulties and delayed marriages, which could mean a return to the bride price practice. Jeffery's discussion on the research literature on marriage and her decades-long fieldwork experience in north-western India convincingly highlight the dangers of demographic determinism and political complacency related to the future of gender relations in India.

The last article by Laila Ashrafun and Minna Säävälä analyses a social innovation – ADR – that has been adopted in some Bangladeshi cities in cases of domestic violence and marital breakdown. ADR is a mediation process involving engaging a neutral arbitrator in disputes to avoid judicial proceedings. This innovation, which originated in the USA, is widely diffused in post-industrialized countries, while USAID and the World Bank, among others, propagate its use in Third World countries. Although the idea is to advance the interests of the women and to make their voices heard in mediation, the authors show how in their case study the outcome has been a cementing of a young, abused wife's subaltern situation in the hierarchical kinship system. The lack of sanctions against the perpetrators of family violence and since the option of going to court is unrealistic, the mediation process provides young wives with very few options to advance their interests. This mediation has become a way of convincing a battered young wife of the need to 'adjust' and to return to what may be a life-threatening situation at home. The social innovation that was meant to empower young maltreated wives has turned out to be an arena of their continued subordination. However, the social innovativeness of the ADR has made it possible to make the difficult situation of young women public, which may have long-term repercussions in society, although even repeated mediations rarely improve individual women's situations permanently.

Conclusion

All these case studies illustrate that innovations do not transform the existing gender and other hierarchies straightforwardly. However, innovation may signal some changes and have effects in the long run. Many of the innovations discussed could have higher transforming potentials if they were offered in a form that allowed women and other subaltern groups to reap more benefits. For example, the dispute mediators in ADR could emphasize young women's preferences more, while offering microloans could be combined with providing women with marketable skills and encouraging them to participate in meetings.

The detailed ethnographic cases in this special issue show the multiplicity of ongoing social processes. Various solutions are often needed, and mere reliance on social innovations and on the non-profit or commercial sectors will not lead to a successful developmental path if these are not backed up with a public sector partnership and networks. Without a functioning judicial system providing back-up, mediation practised in the third sector cannot viably improve the situation of the young wives, and innovations such as microloans cannot compensate for the state's inability to provide social security and health care. Innovations can only be transformative means if they emerge in a juncture where hierarchically lower actors have a pre-existing demand for them and their adoption is combined with other new avenues for action, such as is the new printing technologies that the Urdu press uses. In South Asian societies that are starkly hierarchical and holistic, innovations may have unpredictable repercussions, such as the young women using mobile phones as a means to maintain their support networks with their natal families after marriage. However, an innovation easily ends up serving the interests of the superordinate groups, as has been the case with ultrasound and sex-selective abortions. It is only through meticulous attention to micro data that the impact and nature of innovations can be appropriately assessed and understood.

Acknowledgements

This work was supported by the Academy of Finland [grant number SA138232].

Notes

1. See the special issue on innovation in India (S. Datta 2011).
2. This does not mean that innovations did not emerge before the economic liberalization; however, the pace of innovation seemed to intensify in the more dynamic economic environments.

References

Appadurai, Arjun. 1986. "Is Homo Hierarchicus?" *American Ethnologist* 13 (4): 745–761.

Archambault, Julie Soleil. 2010. "La fièvre des téléphones portables: Un chapitre de la 'success story' mozambicaine?" *Politique Africaine* 117 (1): 83–105.

Arthur, W. B. 2005. "The Logic of Invention." Santa Fe Institute Working Papers 2005-12-045. http://www.santafe.edu/media/workingpapers/05-12-045.pdf

Barendregt, Bart. 2008. "Sex, Cannibals, and the Language of Cool: Indonesian Tales of the Phone and Modernity." *The Information Society* 24 (3): 160–170.

Baruah, Bipasha. 2002. "Challenges for Microcredit in South Asia." *Women and Environments International Magazine* 54/55 (Spring): 27–30.

Basu, Alaka Malwade, and Sajeda Amin. 2000. "Conditioning Factors for Fertility Decline in Bengal: History, Language Identity, and Openness to Innovations." *Population & Development Review* 26 (4): 761–794.

Bellemare, Guy, and Jacques L. Boucher. 2006. "Economie sociale et innovation sociale." *Economie et Solidarités* 37 (1): 4–12.

Bhatt, Ela. 1995. "Women and Development Alternatives: Micro and Small-Scale Enterprise in India." In *Women in Micro and Small-Scale Enterprise Development*, edited by Louise Dignard and Jose Havet, 86–100. Boulder, CO: Westview Press.

Bijker, Wiebe E., Thomas P. Hughes, and Trevor F. Pinch, eds. 1993. *The Social Construction of Technological Systems: New Directions in the Sociology and History of Technology.* Cambridge, MA: MIT Press.

Bijker, Wiebe E., and John Law, eds. 1992. *Shaping Technology, Building Society: Studies in Sociotechnical Change.* Cambridge, MA: MIT Press.

Brown, Louise. 2007. "The Adoption and Implementation of a Service Innovation in a Social Work Setting – A Case Study of Family Group Conferencing in the UK." *Social Policy and Society* 6 (3): 321–332.

Chatterjee, Partha. 1993. *The Nation and its Fragments: Colonial and Postcolonial Histories.* Princeton: Princeton University Press.

Chidambaram, R. 2007. "Global Tour of Innovation Policy – Indian Innovation: Action on Many Fronts." *Issues in Science and Technology* 24 (1): 59–68.

Chhetri, Netra, Pashupati Chaudhary, Puspa Raj Tiwari, and Ram Baran Yadaw. 2012. "Institutional and Technological Innovation: Understanding Agricultural Adaptation to Climate Change in Nepal." *Applied Geography* 33 (1): 142–150.

Datta, Punita Bhatt. 2011. "Exploring the Evolution of a Social Innovation: A Case Study from India." *International Journal of Technology Management & Sustainable Development* 10 (1): 55–75.

Datta, Surja. 2011. "Introduction to the Special Issue on Innovation Dynamics in India." *International Journal of Technology Management & Sustainable Development* 10 (1): 3–9.

Dearing, J. W. 2009. "Applying Diffusion of Innovation Theory to Intervention Development." *Research on Social Work Practice* 19 (5): 503–518.

De Sardan, J. 2005. *Anthropology and Development: Understanding Contemporary Social Change*. London: Zed Books.

Doron, Assa. 2012. "Mobile Persons: Cell Phones, Gender and the Self in North India." *The Asia Pacific Journal of Anthropology* 13 (5): 414–433.

Dumont, Louis. 1970. *Homo Hierarchicus: An Essay in the Caste System*. Chicago: Chicago University Press.

Dror, David M., Ralf Radermacher, Shrikant B. Khadilkar, Petra Schout, François-Xavier Hay, Singh Arbind, and Ruth Koren. 2009. "Microinsurance: Innovations in Low-Cost Health Insurance." *Health Affairs* 28 (6): 1788–1798.

Featherstone, Katie, Paul Atkinson, Aditya Bharadwaj, and Angus Clarke. 2005. *Risky Relations: Family, Kinship and the New Genetics*. London: Berg.

Gupta, Dipankar. 2000. *Interrogating Caste. Understanding Hierarchy and Difference in Indian Society*. Delhi: Penguin Books.

Horst, Heather, and Daniel Miller. 2006. *The Cell Phone: An Anthropology of Communication*. Oxford: Berg.

Jouhki, Jukka. 2013. "A Phone of One's Own? Social Value, Cultural Meaning and Gendered Use of the Mobile Phone in South India." *Journal of the Finnish Anthropological Society* 38 (1): 37–58.

Karim, Lamia. 2008. "Demystifying Micro-Credit: The Grameen Bank, NGOs, and Neoliberalism in Bangladesh." *Cultural Dynamics* 20 (1): 5–29.

Kundu, Amitabh. 2001. "Institutional Innovations for Urban Infrastructural Development: The Indian Scenario." *Development in Practice* 11 (2/3): 174–189.

Mulgan, Geoff. 2007. *Social Innovation: What it is, Why it Matters and How it can be Accelerated*. With Simon Tucker, Rushanara Ali and Ben Sanders. Skoll Centre for Social Entrepreneurship working paper, University of Oxford. London: Young foundation.

Nussbaum, Martha S. 2000. *Women and Human Development: The Capabilities Approach*. Cambridge: Cambridge University Press.

Padmanaban, P., Parvathy Sankara Raman, and Dileep V. Mavalankar. 2009. "Innovations and Challenges in Reducing Maternal Mortality in Tamil Nadu, India." *Journal of Health, Population & Nutrition* 27 (2): 202–219.

Pande, Amrita. 2009. "'It May Be Her Eggs but it's My Blood': Surrogates and Everyday Forms of Kinship in India." *Qualitative Sociology* 32 (4): 379–397.

Parmar, Aradhana. 2003. "Micro-Credit, Empowerment and Agency: Re-Evaluating the Discourse." *Canadian Journal of Development Studies* 24 (3): 461–476.

Patel, Sheela, and Sheridan Bartlett. 2009. "Reflections on Innovation, Assessment, and Social Change: A SPARC Case Study." *Development in Practice* 19 (1): 3–15.

Pinch, Trevor, and Wieber Bijker. 1984. "The Social Construction of Facts and Artefacts: Or how the Sociology of Science and the Sociology of Technology Might Benefit Each Other." *Social Studies of Science* 14 (3): 399–441.

Pol, Eduardo, and Simon Ville. 2009. "Social Innovation: Buzz word or Enduring Term?" *The Journal of Socio-Economics* 38 (6): 878–885.

Rahman, Aminur. 2001. *Women and Microcredit in Rural Bangladesh: An Anthropological Study of Grameen Bank Lending*. Boulder, CO: Westview.

Richman, Barak D., Krishna Udayakumar, Will Mitchell, and Kevin A. Schulman. 2008. "Lessons from India in Organizational Innovation: A Tale of Two Heart Hospitals." *Health Affairs* 27 (5): 1260–1270.

Rogers, E. M. 2003. *Diffusion of Innovations*. 5th ed. New York: Free Press.

Säävälä, Minna. 1999. "Understanding the Prevalence of Female Sterilization in South India." *Studies in Family Planning* 30 (4): 288–301.

Sen, Amartya. 1999. *Development as Freedom*. Oxford: Oxford University Press.

Shove, Elizabeth, Mika Pantzar, and Matt Watson. 2012. *The Dynamics of Social Practice: Everyday Life and how it Changes*. Los Angles: Sage.

Srinivas, M. N. 1989. "Some Reflections on the Nature of Caste Hierarchy." *Contributions to Indian Sociology* 18 (2): 51–67.

Tacchi, Jo, Kathi Kitner, and Kate Crawford. 2012. "Meaningful Mobility: Gender, Development and Mobile Phones." *Feminist Media Studies* 12 (4): 528–537.

Tenhunen, Sirpa. 2008. "Mobile Technology in the Village: ICTs, Culture, and Social Logistics in India." *Journal of the Royal Anthropological Institute* 14 (3): 515–534.

Van de Ven, A. H., D. E. Polley, R. Garud, and S. Venkataraman. 1999. *The Innovation Journey.* New York: Oxford University Press.

Wajcman, Judy. 2002. "Addressing Technological Change: The Challenge to Social Theory." *Current Sociology* 50 (3): 347–363.

Welz, Gisela. 2003. "The Cultural Swirl: Anthropological Perspectives on Innovation." *Global Networks* 3 (3): 255–270.

Williams, Robin, and David Edge. 1996. "The Social Shaping of Technology." *Research Policy* 25 (6): 856–899.

Winslow, Deborah. 2009. "The Village Clay: Recursive Innovations and Community Self-Fashioning among Sinhalese Potters." *Journal of the Royal Anthropological Institute* (N.S.) 15 (2): 254–275.

Wire, Thandika Mkanda. 2007. "Transformative Social Policy and Innovation in developing Countries." *The European Journal of Development Research* 19 (1): 13–29.

Yunus, Muhammad. 2007. *Banker to the Poor: Micro-Lending and the Battle against World Poverty.* New York: PublicAffairs.

Katibs and computers: innovation and ideology in the Urdu newspaper revival

Mark Allen Peterson

Department of Anthropology, Miami University, Oxford, OH, USA

In 1993, the prognosis for Urdu newspapers in north India was dismal. The readership was aging and dwindling as the new generation learned Hindi in Devanagiri script. Urdu calligraphers (*katibs*) were not passing their skills on to a new generation and writers skilled in Urdu were becoming increasingly hard to find. Fifteen years later, Delhi is home to a prosperous and expanding Urdu press. The number of newspapers had tripled, circulations were often higher than they had been in the past, profits were up and the atmosphere at Delhi's major Urdu newspapers was upbeat. A large part of the explanation lies in the intersection of language ideologies and new writing technologies. On the one hand, Urdu indexes crucial politically urgent populations, leading to a renewed interest in it from many sectors. On the other hand, new more flexible technologies allowed the retiring *katibs* to be replaced by computer typesetting that strongly resembles north Indian calligraphic styles and new software allows an entire daily newspaper to be assembled and sent to press using a single laptop. Drawing on ethnographic fieldwork in 1993 and 2008 at some of New Delhi's Urdu dailies and interviews with several editors, this article describes the mutual influences of Urdu language ideologies about Muslim identity and technological innovation in the revival of the Urdu daily.

In December 2008 I sat before an ancient wooden desk in the office of Muzaffar, the circulation director of one of Delhi's oldest Urdu dailies. From a wooden cabinet he brought out a thick pile of Urdu newspapers: *Pratap, Milap, Jadeed In-Dinon, Sahaa Fut, Jadeed Khabar, Hamara Samaj, Inquilaab-e-Hind, Halat-e-Watan, Hindustan Express, Akhbar-e-Mashriq* and *Daily Tej*.

'Today there are maybe 25 Urdu newspapers [in Delhi]', he tells me:

> There is a simple reason. It is cheaper now to put out a newspaper. ... If I have just a staff of four or five, I can put out a newspaper, four or eight pages, and I will have enough ads to make a good profit.

It was a stunning contrast from my previous visit 15 years earlier in 1993. Then, amid a wider boom in vernacular language newspapers (Jeffrey 2000), the prognosis for Urdu newspapers in north India was dismal. The readership was aging and dwindling as the children of New Delhi's Urdu reading communities learned Hindi in Devanagiri script. The Urdu calligraphers (*katibs*) who wrote the newspapers by hand were not passing their skills on to a new generation and writers skilled in Urdu were becoming increasingly hard to find. Delhi had but four Urdu newspapers, none of which claimed a circulation greater than 25,000. Editors (who were often also the publishers) soberly informed me that their newspapers were unlikely to survive them.

Fifteen years later, India is home to a prosperous and expanding Urdu press. Demographically, little had changed: the mean age of readers was over 50 and the katibs had ceased to exist, yet the number of newspapers had sextupled. Delhi had some 25 Urdu newspapers, reported circulations were higher than they had been in the past, profits were up and the atmosphere at Delhi's major Urdu newspapers was upbeat.

In some ways, the revival of the Urdu press can be seen as part of the more general Indian newspaper revival, whose roots go back to the mid-1970s. During this period, the Indian mediascape shifted dramatically from being dominated by a handful of progressive English language newspapers aimed at political, social and economic elites (Peterson 1996; Raghavan 1994, 142–164) to experiencing an explosion of news media in 'vernacular' Indian languages. Jeffrey (1993, 2000) posits five basic reasons for the 140% rise in newspaper circulation between 1976 and 1988: improvements in technology, allowing for more production of more attractive newspapers, as well as electronic transmission to off-site print centers in outlying markets; rising literacy; growing purchasing capacity among consumers; increasing political awareness among many citizens outside the urban centers, fueling a 'desire to know'; and aggressive efforts by publishers to increase their power, prestige and profits through expansion. In the early 1990s, the growth of vernacular news in India received a new boost as the nation embraced economic liberalization. A new boom in advertising revenues boosted growth, in part by allowing newspapers to dramatically cut their prices, relying on ad revenues to cover the costs (Kohli-Khandekar 2006). The rise of cable television as a challenge to the state monopoly on broadcast news did not marginalize print news but actually boosted consumer desire for news, as did changes in local politics that created an entirely new class of local politicians at the rural level (Ninan 2007). New technologies also enabled newspapers to create intimate, local editions and inserts and to reach out to ever more distant potential audiences.

Muzaffar's explanation is part of a general socioeconomic argument that roots the revitalization of Urdu newspapers in the changes wrought by economic liberalization. In a nutshell, this argument states that Urdu is benefiting from the general boom in the news industry wrought by globalization – a gloss at once for the appropriation and adaption of new technologies, the opening up of markets, dropping of governmental economic barriers and rise in consumer spending. These changes have made it possible to run a newspaper for a local community at a very low price. There are more commercial ads than ever before and ad space can be sold for less because of the lower production costs offered by technology.

Muzaffar's explanation is certainly correct – but it is not the whole story. These economic and technological explanations are rooted in a hidden ideological assumption that Urdu is a Muslim language. This paper seeks to complicate the sociotechnical and political economic argument by recognizing the revitalization of the Urdu press as part of a process by which Urdu is being dialectically *created* as the language of India's Muslims. I will argue that the contemporary Urdu news revival is part of a century-old evolving language ideology; that is, 'sets of belief about language articulated by users as a rationalization or justification of perceived language use' (Silverstein 1979, 193) which 'envision and enact links of language to group and personal identity to aesthetics, to morality, and to epistemology' (Woolard and Schieffelin 1994, 55–56). Language ideology has emerged in the last decade as a principle framework for understanding language differentiation, change and maintenance. Language ideologies are frequently contested; they involve cultural struggles over who and what is referred to by particular ways of speaking and writing. In the case of Urdu, it is a struggle over whether Urdu symbolizes a certain level of education and cultural refinement, as it did for most Indians born before the Partition of India in 1947, whether it indexes Muslim identity, as it does for most Indians born after the creation of Pakistan, or whether it indicates a particular lower class, uneducated, conservative Muslim identity, as it does for many Indians, Hindu and Muslim alike, since the rise of strong Hindu nationalist movements in the 1980s.

Global technologies

Muzaffar's account of the small explosion of Urdu newspapers is based on the capacity of information and communications technologies to help small newspapers with relatively low cash flow to manage news gathering, production and distribution with very low investment. One can see this at once by visiting the low-rent office buildings clustered in three or four key locations, from which most Urdu newspapers in Delhi are published. Five or six of New Delhi's Urdu newspapers – maybe 20% – publish from offices in Laxmi Nagar or adjoining Geeta Colony and with good reasons. Rents are low and the location is ideal. The inexpensive printing presses in Noida where most Urdu papers publish are just 10 minutes south, while 10 minutes west across the Yamuna river is Bahadurshah Zafar Marg, 'the Fleet Street of Delhi' home of both the great English and Hindi dailies like *Times of India*, *Indian Express*, *Nav Bharat Times* and *Business Standard* but also of the once great but declining Urdu giants *Milap*, *Pratap* and *Tej*. Turn north and a five-minute drive brings you to Daryaganj, center of New Delhi's book publishing industry and home to another cluster of Urdu newspapers, tiny four pagers often serving only local communities clustered around some mosque or ancient Muslim neighborhood, such as *Al-Jamait*, *Views Times*, *Akhbar-e-Nav* and *Din Dunia*.

Changes in information and communication technologies (ICT) have revolutionized Urdu newsmaking in at least three distinct ways. First, ICT have transformed the gathering of news. Second, India has become a manufacturer of newspaper presses, rather than a consumer. Third, ICT have revolutionized newspaper distribution.

Perhaps 80% of all Urdu news comes from the government sponsored Urdu service of the Press Trust of India (PTI), according to PTI's own estimates. PTI is the largest of India's news services (the other is the declining United News of India but there are also a growing number of small, specialized private news services, including one small Urdu service). Although initially designed for older teletype delivery, PTI and other services now usually deliver directly to computers in digital formats which can be immediately edited and pasted into newspapers.

Maulana Wahiduddin Khan, one of the few writers to clearly distinguish between the Urdu generally and the specifically 'Muslim' press, once lamented the parochial nature of India's Muslim press and its failure to attend to, much less contribute to, news of even regional, much less national and international, interest (Khan 1994, 70). In fact, it is detailed attention to local community interests that has increasingly allowed newspapers in India to weather the emergence of new media like private television news and the Internet.

News feeds that flow directly into the laptop allow editors to quickly read, edit and incorporate regional, national and local news they believe would be of interest to their community of readers quickly. A content analysis of four leading Urdu dailies suggests that a little more than 70% of this news comes directly from the PTI Urdu service, a figure very close to PTI's own estimates. However, at newspapers like *Jadeed Khabar*, which pride themselves on a more international and global outlook, Internet technologies also enable coverage of global events the editor believes would interest local communities. Via the Internet, reporters can rapidly accumulate information about events occurring in other parts of the world and write a story original to the newspaper from facts gleaned from these sources.

Once the newspaper is laid out electronically, the files can be transmitted electronically to one of many printing presses available at a low cost. Until the turn of the millennium, acquisition of a press was a major limiting factor on newspaper growth. Presses had to be imported from Europe, usually Germany, and paid for in European currency. Moreover, press ownership required a staff of trained operators. Presses were thus prohibitively expensive for small newspapers and credit was difficult to leverage.[1] Small Urdu newspapers were usually dependent on very old, small presses or paid larger newspapers to print their papers during times the press was not in use. Some of the older Urdu newspapers like *Pratap* helped fund themselves during periods of decline by renting out press time.

The decline in newspapers in North America and Europe made press manufacturing increasingly less profitable in the West, however, and in the late 1980s, Indian companies like Manugraph (Mumbai), Pressline (New Delhi) and The Printer's House (in Faridabad) licensed German technologies and began manufacturing sophisticated, computerized modular presses. Locally manufactured, sold and distributed presses became far more available and less costly. While most Urdu newspapers still cannot afford a press, the sheer number of presses available in India, and the sophistication of the technology, brings publication costs down to surprisingly manageable levels. Many new publishing companies sprang up in the rising media city of Noida across the Yamuna River from Delhi, where a once working-class cluster of villages was being transformed by new media companies and upscale high rises for middle-class Delhi commuters.

Meanwhile, back in Delhi, the Suri brothers, heirs of *Milap*, one of the oldest surviving Urdu dailies, gestured proudly at a MacIntosh laptop. 'The entire newspaper can be run from here', said Yogendra. The eight-page newspaper is edited on a laptop and Yogendra keeps his distribution data on another laptop. Database technologies and spreadsheets allow *Milap* to track every subscriber. Combined with an improved transportation infrastructure and mobile phone technology, *Milap* can serve ever smaller constituencies. The distribution center locations have come to rely on sophisticated software-driven systems to manage the age-old bicycle distribution system for newspapers in New Delhi and the surrounding areas. Combined with changing transportation infrastructures, the new technologies allow the newspaper to distribute in a timely way even to small villages 'with only one or two subscribers', a feat that was inconceivable just a decade earlier.

The capacity of new technologies to cut costs, to replace skilled labor and to increase efficiencies has allowed declining newspapers like *Pratap* and *Milap* to maintain themselves and allowed more than a dozen rivals to emerge. Yet this takes place within a wider context than merely that of neoliberal globalization. There is more advertising because business is booming, but why do advertisers want to put their advertising in tiny Urdu newspapers? Computers let newspapers search, download, edit and compose news stories from the Urdu PTI service, but why does such a government-subsidized resource exist, making stories available at prices that these small newspapers can afford? A partner in a private Urdu press service supplying television as well as print news admitted that his company draws more revenue from clients paying to place news – especially foreign governments ranging from the Gulf states to the USA – than from news media clients who receive the news. It is only in the context of the growing understanding of Urdu as the language of India's Muslims that these make sense.

Making Urdu

'We are not interested in writing "pure" Urdu but in reaching as many people as possible', said Mr Mutawali, editor of a small Urdu daily in Delhi, expressing a sentiment I subsequently heard several times:

> If the people on the street use a Hindi word, we use that. If the person on the street typically uses an Urdu word, we use that word. Even English, if most people are preferring an English word, we will use that.

To speak of Urdu and Hindi – to make of them objects of discussion, analysis and assessment – is to enter into a complex dialectical debate because the two languages define themselves to a large extent in contradistinction to one another.

Syntactically, spoken Hindi and spoken Urdu are so similar that most linguists describe them as constituting a single language continuum, although their phonology, morphology and vocabulary differ in some respects. It is still common in many places to speak of a single spoken language, *Khari Boli* (literally, 'standard speech', usually called Hindustani in English), of which Urdu and Hindi are written variants. Yet so fraught are the language ideologies that define Hindi and Urdu that to speak of a common spoken language or to refer to a past in which a single language served as a lingua franca is itself to enter into an ideological debate, marshaling complex, ambiguous and sometimes contradictory evidence in support of historical narratives that serve particular nationalist (pluralist, Hindutva or other) agendas.

If we refer to the spoken continuum of north Indian language as Khari Boli, it refers to a speech resource rich with lexical possibilities, since for every 'Urdu' word of Persian or Arabic derivation there is a corresponding Hindi word of Sanskrit derivation. In language use, the value of one's choice of one term over the other can have significant performative consequences. A large number of words are simply 'shared' – everyone uses them and no one considers their origins. In everyday speech, no one ever says the Hindi *pustak* (derived from Sanskrit) for 'book' over the Urdu *kitab* (derived from Arabic via Persian); to do so could only be intended to mark some political or other distinction. Many lexical terms are intentionally used as indicators that one is speaking 'Urdu' instead of 'Hindi', especially words containing the phonemes /f/, /z/, /kh/, /gh/ and /q/. Because the language of the popular Hindi film is perhaps the purest expression of contemporary Khari Boli,[2] the 2008 hit film *Jodaa Akbar* offers a particularly interesting example of how the two languages can be articulated in spoken performance. Many filmgoers remarked to me how the Hindu princess, played by Aishwarya Rai, speaks 'pure' Hindi, while Mughal Prince Akbar, played by Hrithik Roshan, speaks 'pure' Urdu, without noting the irony that audiences can understand both fluently and tell the difference.

In understanding this difference-making process, it is useful, perhaps, to draw on the distinction between 'code-switching' and 'switching codes' (Alvarez Caccamo 1998). Both insertional and alternational code-switchings are commonplace in everyday language across northern India. Insertional code-switching – not only Hindi/Urdu but increasingly English – certainly indicates a 'mixed code' (Auer 1998).[3] Certain forms of code-switching, however, serve as metamessages to indicate that one is switching codes – speaking specifically Hindi or Urdu.

It is in writing practices that the 'difference-in-sameness' (Bard and Ritter 2008, 21) that characterizes the continuum of spoken northern Indian language is most fully transformed into linguistic difference. Thus while 'at the colloquial level, in terms of grammar and core vocabulary, [Hindi and Urdu] are virtually identical. … At formal and literary levels … vocabulary [differs] … to the point where the two languages/styles become mutually unintelligible' (Masica 1993, 27). Literary Hindi, in other words, is produced by rendering the Indic grammar of Khari Boli in Devanagari script and introducing Sanskritized elements, while literary Urdu is produced by rendering the standard speech into Urdu script and introducing Persio-Arabic elements. That this is a continuum in which conscious authorial choices must continually be made is emphasized by the fact that Devanagiri script differs from Sanskrit in part by the introduction of several orthographic characters needed to render Persio-Arabic sounds that do not occur in the pre-Islamic language and Urdu script differs from its Persio-Arabic original in its possession of characters that express Indic sounds not present in Arabic or Persian. Thus, particular lexical and idiomatic choices create language forms that frame active understanding, reception and evaluation of content. The use of these characters in either language indicates the use of a word from the opposite language.

The acts of writing through which Urdu newspapers are produced are hence always performances, actions that situate the writers within a complex sociolinguistic field, one fraught with political issues. When an editor chooses to emphasize 'high' Urdu, with Persianized language and poetic registers, he ties the paper primarily to the dwindling generation of readers for whom Urdu is not the language of the Muslim masses while increasing the sense of alienness experienced by other Indians. If the language of the newspaper follows the 'common language of the street', the newspaper risks indexing the uneducated Muslim masses constructed by the rhetoric of Hindu nationalists. To write Urdu is thus always to commit a political act; there are no neutral registers.

This is especially the case when one speaks, as does Mr Mutawali, of reaching the Urdu 'masses' a group that exists primarily in the imaginations of Muslim and Hindu political leaders, but increasingly also in the rhetoric of Urdu newspaper editors. When a mixed group of Indians gathers in a metropolis like Delhi for a get-together, talk often turns at some point to a discussion of regionalism. People will speak of the 'sharpness' of their Bengali spouse, or tell stories about the difficulties in their marriage because his manners and customs were shaped by his soft-spoken Haryana background while she is a spicy Punjabi girl or even discuss whether their kids are more Marathi like their mother or more Gujarati like their Dad. Such stories reflect an important 'regionalism' in the construction of identities which is indexed through language.

By contrast, one of the ways in which Muslims are constructed as an internal Other within the Indian state is by de-regionalizing them. In popular films, Muslims are often presented not as Bengalis, Haryanans, Punjabis, Gujaratis and Marathis, indexed as such by their language, dress and behavior but as 'generic' Muslims, speaking Urdu, women wearing the veil and men carrying the *tashbih* (prayer beads) (Dar 2008). In fact, Muslims for whom Urdu is the mother tongue are increasingly rare.[4] Yet Urdu *is* a Muslim language and increasingly so as the last generation of Hindu partition migrants dies out. Regional Indians in West Bengal, Tamil Nadu, Andhra Pradesh, Punjab and elsewhere throughout India learn literary Urdu as a second, third or even fourth language in madrasas and specialized Urdu schools (Ahmed 2002; Benei 2008; Matthews 2003; Metcalf 2010). In the madrasas, Urdu serves as an important 'bridge language' to Arabic, as one Urdu teacher explained to me. The orthography can point them to Arabic, he said, while the familiarity of the grammar and lexicon is reassuring.[5]

How Urdu becomes Muslim

The terms Hindi and Urdu themselves date back to the Mughal court period, when Hindi, derived from an ancient term for the Indus river, was used to designate anything local and 'Indian'. Urdu derives from a Turkic word meaning 'army camp' within which the language was said to have developed as a kind of lingua franca (Shackle and Snell 1990). By the four-teenth century, Urdu was well established as the canonical literary language of 'high culture' as far south as the Deccan. The Muslim kingdoms of the era often supported ethnic, linguistic and religious pluralism and court poets performed devotional poetry to Krishna in pre-modern north Indian dialects like Braj Bhasa, while Urdu poets referred to their own highly Persianized literary tradition as 'rekhta' (mixed) distinguishing it both from the Persian of the court administration and the colloquial Urdu that 'smacked of the bazaar and rough, uncultured armies' (Saksena 1996, 7).

Urdu enjoyed a significant revival as an intellectual language during the mid- to late nineteenth century when Christian missionaries proselytizing Hindus and Muslims in the Punjab and other northern regions spawned an intellectual effort toward 'reform and rede-finition within the community as a means to strengthening it and enhancing its position in the country' (Dulai 1982, 176). These religious reform movements coincided with and helped shape various forms of nationalism, regional language movements and agitation for vernacular education throughout India. Nationalist and colonialist projects ironically worked together to privilege Urdu, with its broad audience, over local languages like Punjabi and Braj Bhasha. Over the course of the next half century, though, this:

> Hindustan lingua franca, written in either script (although Urdu script was dominant), and without regard to any politics of etymological provenance, underwent a transformation in the late nineteenth century. 'Hindi' became associated with the classical and ritual language of Sanskrit, the Devanagiri script, as well as the variety of religious practices known as 'Hinduism,' in contradistinction to 'Urdu,' associated with Persian, Arabic, a modified Persian script, and Islam. (Bard and Ritter 2008, 25)

As British power grew in the eighteenth and nineteenth centuries, Urdu was 'utilized as a language of power complicit in the formation of a new service gentry loyal to the British and subservient to the British way of life in the nineteenth century' (Yaqin 2008, 118). Under British patronage Urdu increasingly became a key signifier of Muslim separateness and political separatism (Robinson 2008).

Although Urdu newspapers were political in orientation from their start in the mid-nineteenth century, poetry also played an important role. Many of the greatest Urdu poets were first published in newspapers; some continued to publish their work in newspapers for their entire lives. Of course, much of the poetry was itself overtly or allegorically political. In the context of the rising pro-Western movements articulated by such intellectuals as Sayyid Ahmad Khan, Mohammed Iqbal's poems spoke to the problem articulated by so many colo-nial and postcolonial intellectuals: how to best blend East and West, traditionalism and mod-ernism and love and intellect (Malik 1967). A generation later, amid proletarian uprisings spawned by the great depression and Britain's harsh reprisals, Ahmad Faiz wrote of suffering and revolution; still later his poetry articulated the philosophy of Muslim nationalism (Hashmi and Gardesi 1982). Even as it increasingly is used as a language of undifferentiated, regional masses, Urdu retains the prestige of being connected to forms of lyric poetry often held as among the highest forms of literary expression in any Indic language.

As even this brief overview suggests, Urdu exists within a complex signifying structure. It is the native tongue and literature of hundreds of thousands of aging Hindus who fled to

India during partition, yet it is constructed as the language of South Asian Islam. It is an important literary language for hundreds of thousands of Indian Muslims who speak in everyday life the regional dialects in which they were raised and which they share with their Hindu, Christian, Sikh and Parsee compatriots. It is the national language of Pakistan, whose rationale for coming into existence is that Muslims constitute a separate South Asian nation (*qaum*) and therefore required a separate state (*watan*) and whose political sundering from India remains the central historical trauma for both nations. Urdu is thus heavily involved in the complex dynamics through which the construction of the Muslim 'Other' within India is tied to the dialectics of the India–Pakistan relationship and the feelings of betrayal and disdain for Pakistan expressed by many Indians, Muslim and Hindu (Dar 2008). I want to follow Rizwan Ahmad in arguing that 'the Urdu language represents a palimpsest of indexicality – layers of meaning deposited one over another [so that] ... the sociolinguistic field of Old Delhi is marked by different, often competing perceptions of Urdu' (Ahmad 2008, 1).

Urdu newspapers position themselves within this sociolinguistic field, inviting readers to participate in a particular approach to the world expressed in a particular language. Readers in Delhi, with more than a dozen English daily newspapers, twice as many Hindi newspapers and at least 25 Urdu dailies, have a large surplus of newspapers to choose from, so that which newspaper one chooses to read indicates particular categorizations of person (Peterson 2009).

The language ideology that defines Urdu as a Muslim language is rooted in a classic sociolinguistic system of reference in which Urdu orthography indicates Muslim identity. This is rooted in a deeper referentiality in which Urdu conjures images of madrasa schools, media stereotypes of veiled women and bead-carrying men, the historical Muslim separatist movement, but most of all Pakistan, the frightening neighbor ripped from the very body of India, with whom three wars have been fought and whose existence is justified by a political theory that equates religious community with national identity. Yet such references are themselves constructed through the careful attention to certain histories and geographies and by careful inattention to alternative, but equally true, histories and demographics. It requires suppression, for example, of the fact that Urdu continues to be the native tongue of more Hindus than Muslims in India and that Urdu is not the native tongue of most Pakistanis. It requires the great catastrophe of partition to be placed solely on the shoulders of Jinnah, who advocated for Urdu as Pakistan's national language, for the important roles of many other separatist leaders speaking Bengali and other regional languages to be suppressed, along with the role of Hindu nationalist leaders for whom partition 'greatly simplified' the political process, as Nehru himself observed.

Three of the oldest surviving Urdu newspapers in Delhi are not 'Muslim' newspapers. The majority of their readers are, and always have been, Hindus. *Milap*, founded in Lahore 1923 as an Arya Samaj organ, takes a strongly secular nationalist stance that insists on the importance of communal harmony. *Pratap*, also founded as an Arya Samaj organ in 1919, has for decades taken an increasingly strident Hindu nationalist line. Both newspapers came to Delhi during Partition in 1947, after the seizure of their properties and following the exodus of their predominantly Hindu readership. As well-established newspapers they received places on Bahadur Shah Zafar Marg, the 'Fleet Street' of India. They were joined by a third Arya Samaj newspaper, *Hind Samachar*, headquartered in Jalandhar, Punjab but also publishing a Delhi edition. All three played significant roles in Indian politics in the 1950s and 1960s as powerful voices for the Urdu speaking post-partition community in Delhi.

By the late 1970s, though, things were changing. Urdu increasingly was coming to be seen as the language of India's substantive Muslim minority. At this time the government established its own 'official' Urdu newspaper for Muslims, *Qaumi Awaz*. *Pratap*, *Milap* and *Hind Samachar* began to lose advertising revenue as their circulations declined with changing audience demographics – older Urdu literate families who had come to India during partition were raising their children as Hindi-literates. The newspaper companies began to diversify. Jagat Narain, founder of *Hind Samachar*, established a Hindi daily, *Punjab Kesari*, *Milap* created new editions in Hindi and the Narayan family, editors and publishers of *Pratap*, began the Hindi daily *Vir Arjun*. These routine business practices of adapting to changing market conditions are also cultural processes that forward the language ideology linking Urdu to Muslims. As the population for whom Urdu represents poetry, refinement and education dwindles and does not reproduce itself, it is subject to an erasure. As this audience, with its public counter-voice dwindles, Urdu script increasingly iconizes Muslims, though its visual similarities to the foreign Persian and Arabic scripts. The orthographic distinctions between Hindi and Urdu become recursively mapped onto the new newspapers in Hindi founded by these two companies as a way to keep the companies alive.

The vanishing of the Katib

One of the clearest examples of the ways technologies and language ideologies work together to shape social change can be found in the disappearance of the professional Urdu calligrapher or *katib*. To an outsider, the only clear distinction between Urdu and Hindi is orthographic. Urdu employs a Persian script, itself a modification of the Arabic script, with special graphemes for such native Indic phonemes as aspirates. As in Persian and Arabic, short vowels are not written except in specialized texts such as dictionaries and textbooks. Distinctions between words whose meaning is determined by an unwritten short vowel, such as /pƏl/ 'moment', /pIl/ 'to labor' and /pƱl/ 'bridge', must in most documents be determined from the discursive context in which it is embedded. Urdu flows from right to left with spacing and sometimes even the forms of letters shaped by the style of the writer.

As with Arabic and Persian, Urdu calligraphy is considered an art form and the way Urdu is expressed can be as important to many readers as the content. For centuries, the elegance of the form of Urdu writing was expected by discerning readers to match the significance of the content. In the days before print, writers of Urdu – Muslims and Hindus alike – were expected to have elegant handwriting and the more elegant the expression of thought – especially poetry – the more elegant the calligraphy. For the calligraphically impaired, there arose a class of katibs, Urdu secretaries, who could render one's words into properly elegant form. After the British colonial administration shifted the official language of North Indian courts from Persian to Urdu in 1837, the importance of the katib rose, schools for katibs multiplied and the job became a standard one employed across professions.

The expectation by readers that important Urdu would be expressed in elegant calligraphy meant that as Urdu newspapers began to appear in the mid-nineteenth century they could not employ the movable types used in English newspapers because this would compromise the connections between the letters and inhibit the elegance of form required for the expression of lofty ideas in Urdu. Instead, Urdu newspapers seized on lithographic technologies, already in widespread use in Turkey, Persia and Arabia. Using these technologies, Urdu newspapers could be handwritten and still mass produced.

One of the central figures of Urdu news, then, became the katib, now referring to the scribe who hand wrote every piece of text that went into the newspaper. In 1992, the publisher of *Pratap* gave me a tour of the newspaper. In a large room in the rear of the building were seated around dozen men, on benches at desks, writing with a flourish.

It was an exhausting enterprise. 'The average katib can only produce a column and a half' by publication time, recalled Yogendra Suri, one of the current editors of *Milap*. An 8-page broadsheet newspaper with 6 columns per day thus required about 32 katibs to successfully lay out the newspaper. Katibs arrived in the morning to begin laying in ads and setting the letters, poetry and other non-perishable items. News stories usually had to be handed in or phoned in by reporters by noon to make the 3 pm printing deadlines. Katibs were given pre-eminence on the basis of speed, compactness and elegance of handwriting, with the best calligraphy being reserved for the editorial, since the emphasis in the Urdu newspapers has always been editorial and literary more than news reporting, as several Urdu editors remarked to me.

Since news stories had to be written in around the pre-existing advertising and feature text, the katibs often made judgment calls about how to abbreviate them to fit the available space. One katib told me he would occasionally vowel a word if he felt that its meaning was not clear from the context. Katibs thus also functioned as proof readers and subeditors, correcting grammar or putting awkward phrases into more elegant prose. Occasionally, from the ranks of the katibs would arise a reporter or editor.

This confusion of activity led to a series of labor struggles. Katibs were paid less than half the amount journalists were and were one of several kinds of press workers not covered by the Press Council's strictures on wages for working journalists. After the national journalists' union split in the early 1970s, both unions began agitating to reclassify some unrecognized types of press workers as journalists in order to increase their respective memberships. One argument they made was that if katibs were altering copy rather than merely reproducing it, then they were journalists and subject to wage board strictures. Publishers fought back, with *Pratap* leading the fight. Most of the battles took the form of court cases; no one at the newspapers could recall any strikes, walk-outs or agitations.

Although the publishers usually won the cases, these were Pyrrhic victories because they led directly to the demise of the katib as a class of worker. Katibs were skilled workers who required an extensive knowledge of calligraphy, Urdu literature and grammar. As their low wages failed to keep pace with inflation, fewer and fewer young men apprenticed as katibs. Increasingly, those with the skills to work as katibs were gaining these skills in private Urdu schools or even college and the wages earned by katibs were insufficient to lure these graduates to the job in sufficient numbers, since their educations qualified them for careers with much higher salaries as teachers, secretaries and similar positions.

By the late 1980s, the advent of computer typesetting promised to revolutionize Urdu printing.[6] Urdu fonts were first used in Pakistan, then adopted by Urdu dailies in Hyderabad and only then were they adopted in the north. *Milap* became the first Delhi Urdu newspaper to use computer generated text, but immediately ran into a problem: the fonts, designed to be highly readable, were not orthographically sophisticated enough for *Milap*'s readers. Editor Navin Suri told me in 1993 that it took three tries to finally locate a font that resembled calligraphy closely enough that it could meet readers' expectations. *Milap* continued to employ katibs as headline writers for many years, until the last one retired.

'Today it would be impossible to operate the newspaper with katibs', Suri said when I met him again in 2008. 'Even if one could afford to pay them, where would you find them? They did not pass the art on to their children. There are no more katibs.'

In this relatively straightforward economic process of skilled workers replaced by technologies, we also see an expression and realization of a language ideology. The writing of the katibs iconizes the conceptualization of Urdu as a language of elite refinement. Their disappearance from the news business is replicated at the level of readership communities, with the decline of Partition-era Urdu readers who insist on the elegance of calligraphic Urdu. In the process, the belief in Urdu as a pan-Indian language is rendered increasingly tenuous and the political ideology which links Urdu to an imagined generalized Muslim population is strengthened.

Conclusion

Since India began to revise its market economy to be more in line with global neoliberal practices in the early 1990s, it has been filled with stories of entrepreneurial success, including in the media. In these tales, media successes – measured by rising audiences, ad revenues and profits – are a direct result of the changing economy of investment, credit capital, deregulation and increasing technological efficiency. This paper is part of a larger project in which I seek to challenge the simple technological and economic determinist explanations for social and cultural change not by contradicting them but by emphasizing the significance of local cultural contexts. The current boom in Urdu newspapers cannot be entirely explained either by economic liberalization or by new technologies of reproduction. The revitalization of the Urdu press is a result of the intersection of these processes with shifting language ideologies and new writing technologies in a context of changing attitudes about the communities being indexed by Urdu.

The rise in numbers of Urdu newspapers and numbers of subscribers in New Delhi is predicated not on the regionalism that is central to the broader Indian newspaper upsurge, but on the implementation of new writing practices that allow printing and distribution at low costs, the growing availability of advertising and a state-subsidized Urdu news service. Each of these factors is in turn predicated on the role of Urdu as a pan-Indian language of Muslims. The success of the Urdu press thus introduces a cultural paradox. To produce and market an Urdu newspaper requires Urdu journalists to construct as an ideal readership a delocalized and generalized Indian Muslim population – the very language ideology for defining Urdu put forth by Hindu nationalists and which the proprietors of these newspapers claim to want to undermine. The innovations in India's new economy must thus be seen not as driven generally by market forces, deregulation and new technologies, but also by historical trends specific to Urdu, and which have been decades in the making, including the changing meanings of Muslim identity in India, and by the contingencies of national political concerns.

Notes

1. At least one Indian company, The Printer's House, began manufacturing offset presses in the 1960s. However, the company's presses were often held to be inferior to foreign-manufactured presses and The Printer's House made most of its income importing presses until the 1980s, when it introduced its Orient line of presses based on leased German technology. Even these were out of the price range for most Urdu newspapers, however.
2. Indeed, Kesavan goes so far as to suggest that India's socially realistic films often distance themselves 'from the commercial cinema and its bazaari Urdu' (Kesavan 1994, 251)
3. The distinction between insertions of words of a different code and 'borrowings' can often be determined from transformations of the words in different contexts of use. Thus one might

well hear '*koi* books *shooka rakhe hai*' or '*koi booka shooka rakhe hai*' (do you have any books) in which the former is an insertative code-switch, but the latter – in which the English word is localized by giving it a Hindi/Urdu plural suffix – is a borrowing.

4. Ironically, the use of Urdu in processes of deregionalization and construction of the Muslim Other within India is paralleled in Pakistan, where rich regional languages and the literatures in Sindhi, Punjabi, Balochi and Pashtu (languages which have large communities of speakers across the border in India) are suppressed and marginalized in favor of a single Urdu national culture (Riaz 1986).

5. Since the 1990s, the Indian state has supported this effort with free Urdu educational materials prepared by the Ministry of Human Resource Development (Metcalf 2010) apparently as a way to get Indian Muslims to frame their identity through an Indian language rather than a foreign one such as Arabic or Persian.

6. Earlier, a few Urdu newspapers tried to take advantage of the development of Urdu linotype machines (*chapay ki masheen*) in the 1970s, but the *Nasakh*-style fonts proved unpopular with readers, even when headlines continued to be handwritten, and most returned to lithography.

References

Ahmed, Imtiaz. 2002. "Urdu and Madrasa Education." *Economic and Political Weekly* 37 (24): 2285–2287.

Ahmad, Rizwan. 2008. "Unpacking Indexicality: Urdu in India." Paper presented at the Texas Linguistic Forum, Austin, TX. Accessed January 31, 2012. www.studentorgs.utexas.edu/salsa/proceedings/2008/ahmad_2008.pdf.

Alvarez Caccamo, C. 1998. "From 'Switching Code' to 'Code Switching'." In *Code-Switching in Conversation*, edited by Peter Auer, 29–48. London: Routledge.

Auer, Peter. 1998. "Introduction: Revisiting Bilingual Conversation." In *Code-Switching in Conversation*, edited by Peter Auer, 1–24. London: Routledge.

Bard, Amy, and Valerie Ritter. 2008. "A House Overturned: A Classical Urdu Lament in Braj Bhasha." In *Shared Idioms, Sacred Symbol, and the Articulation of Identities in South Asia*, edited by Kelly Pemberton and Michael Nijhawan, 21–53. London: Routledge.

Benei, Veronique. 2008. *Schooling Passions: Nation, History and Language in Contemporary Western India*. Stanford, CA: Stanford University Press.

Dar, Huma. 2008. "Can a Muslim Be an Indian and Not a Traitor or a Terrorist?" In *Shared Idioms, Sacred Symbol, and the Articulation of Identities in South Asia*, edited by Kelly Pemberton and Michael Nijhawan, 97–114. London: Routledge.

Dulai, Surjit Singh. 1982. "Punjabi: Language Intelligensia and Social Change." In *South Asian Intellectuals and Social Change*, edited by Yogendra Malik, 160–189. New Delhi: Heritage.

Hashmi, Bilal, and Hasan nawaz Gardesi. 1982. "Urdu: The Structural and Cultural Context of Intellectualse." In *South Asian Intellectuals and Social Change*, edited by Yogendra Malik, 202–234. New Delhi: Heritage.

Jeffrey, Robin. 1993. "Indian-Language Newspapers and Why They Grow." *Economic and Political Weekly* 28 (38): 2004–2011.

Jeffrey, Robin. 2000. *India's Newspaper Revolution: Capitalism, Politics and the Indian Language Press, 1977–1999*. Oxford: Oxford University Press.

Kesavan, Muluk. 1994. "Urdu, Awadh, and the Tawaif." In *Forging Identities: Gender, Communities and the State in India*, edited by Zoya Hasan, 244–257. Boulder, CO: Westview.

Khan, Maulana Wahiduddin. 1994. *Indian Muslims: The Need for a Positive Outlook*. New Delhi: Al-Risala.

Kohli-Khandekar, Vanita. 2006. *The Indian Media Business*. 2nd ed. New Delhi: Sage.

Malik, Hafeez. 1967. "The Marxist Literary Movement in India and Pakistan." *Journal of Asian Studies* 17 (4): 649–664.

Masica, Colin. 1993. *Indo-Aryan Languages*. Cambridge: Cambridge University Press.

Matthews, David J. 2003. "Urdu Language and Education in India." *Social Scientist* 31 (5/6): 57–72.

Metcalf, Barbara. 2010. "Madrasas and Minorities in Secular India." In *Schooling Islam: The Culture and Politics of Modern Muslim Education*, edited by Robert W. Hefner and Muhammad Qasim Zaman, 87–106. Princeton, NJ: Princeton University Press.

Ninan, Sevanti. 2007. *Headlines from the Heartland: Reinventing the Hindi Public Sphere*. New Delhi: Sage Books.

Peterson, Mark Allen. 1996. "Writing the News in India: Press, Politics and Symbolic Power." PhD diss., Brown University.

Peterson, Mark Allen. 2009. "Getting the News in New Delhi." In *The Anthropology of News and Journalism*, edited by Elizabeth Bird, 267–288. Bloomington: Indiana University Press.

Raghavan, G. N. S. 1994. *The Press in India*. Delhi: Gyan Publishing House.

Riaz, Fahmida. 1986. *Pakistan: Literature and Society*. New Delhi: Patriot Publishers.

Robinson, Frances. 2008. *Seperatism among Indian Muslims: The Politics of the United Provinces' Muslims 1860–1923*. 2nd ed. Cambridge: Cambridge University Press.

Saksena, Ram Babu. 1996. *A History of Urdu Literature*. Lahore: Sang-e-Meel.

Shackle, Christopher, and Rupert Snell. 1990. *Hindi and Urdu Since 1800: A Common Reader*. London: University of London/SOAS.

Silverstein, Michael. 1979. "Language Structure and Linguistic Ideology." In *The Elements: A Parasession on Linguistic Units and Levels*, edited by R. Cline, W. Hanks, and C. Hofbauer, 193–247. Chicago, IL: Chicago Linguistic Society.

Woolard, Kathryn, and Bambi Schieffelin. 1994. "Language Ideology." *Annual Review of Anthropology* 23: 55–82.

Yaqin, Amina. 2008. "Variants of Cultural Nationalism in Pakistan." In *Shared Idioms, Sacred Symbol, and the Articulation of Identities in South Asia*, edited by Kelly Pemberton and Michael Nijhawan, 115–139. London: Routledge.

Microcredit and building social capital in rural Bangladesh – drawing the uneasy link

Mohammad Jasim Uddin[a,b]

[a]Department of Social Research, University of Helsinki, Helsinki, Finland; [b]Department of Sociology, Shahjalal University of Science & Technology, Sylhet, Bangladesh

In the 1990s, social capital and the group-based microcredit programme emerged as major planks of developmental interventions, and both approaches have underscored the necessity to mobilize social factors in the alleviation of poverty and social solidarity. The group-based microcredit model is considered an effective policy instrument for increasing women's access to financial capital and for strengthening their social capital at the local level. This study contributes to the continuing debate over how or if group-based microcredit facilitates to the formation of social capital at the local level in Bangladesh. Case studies and ethnographic (in-depth) interviews of 151 women microcredit borrowers of the Grameen Bank and the Bangladesh Rural Advancement Committee of Bangladesh were used in this study. The study suggests that the relationship between participation in the group-based microcredit programme and the facilitation of social capital at the local level is at best ambiguous. The assumed association between microcredit membership and building social capital (social networks, norms of reciprocity and collective identity and action) is much less prominent than commonly suggested by many previous scholars and development practitioners.

Background

Providing small amounts of credit at a heavily subsidized rate of interest to rural poor farmers has been put forward as a policy for poverty reduction. However, established financial institutions have failed to offer this service in many developing countries (Adams, Graham, and Pischke 1984; Hoff and Stiglitz 1990). Many of the earlier state-driven financial programmes had weak incentive structures, were heavily bureaucratic, politicized and wrongly chosen (Adams, Graham, and Pischke 1984; Sobhan 1998; Bastelaer 2000). The bureaucrats and local wealthy and powerful elites siphoned off subsidized aid money meant for the poor (Adams and Vogel 1986; McGregor 1988). Unlike in industrial economies, the most economically disadvantaged and vulnerable people in developing countries often do not have access to financial organizations and are regarded as credit ineligible because they cannot put up the collateral that is acceptable to formal banking organizations. Dr Muhammad Yunus first launched a systematic 'group-based'[1] microcredit programme in Bangladesh through the Grameen Bank (GB) in the mid-1970s in reaction to these shortcomings of the formal banking systems and to alter the lives of the world's poorest. The bank earned an almost 'mythical reputation' among aid donors and the policy-makers as a panacea for alleviating poverty, particularly among women. In the 1990s, the World Bank and other donor agencies indeed touted the microcredit programme as a central policy intervention for poverty alleviation that helps the poor's access to financial services and contributes to gender empowerment in developing countries (Binswanger and Landell-Mills 1995; Mayoux 2001, 2002).

Almost simultaneously, social scientists and development practitioners noticed that the forms of capital conventionally used in development theory (natural, physical and human) were failing to spot an important element that is generally known as social capital (networks of relationships, norms, collective identity and action), even though the concept is about a century old (Bastelaer 2000). Group-based microcredit programmes are considered as effective policy instruments for creating and strengthening the social capital of the community. In order to reach as many of the clients as possible, microcredit programmes provide credit using social mechanisms such as self-selected group formation, the social collateral mechanism or joint liability, and a regular presence at weekly instalment group meetings (Stiglitz, 1990; Holt and Ribe, 1991; Yaron, 1994; Larance, 1998; Anderson, Locker Leigh, and Nugent 2002; Dowla, 2006). Professor Yunus also declared that the system of microcredit values 'gives high priority on building social capital' (Woodworth, 2008, p. 36). The main objective of this paper is to explore whether women's participation in 'group-based' micro-lending programmes reinforces and facilitates social capital at the village level in Bangladesh. The study is based on the research I conducted among micro-credit borrowers of the GB and the Bangladesh Rural Advancement Committee (BRAC) in Sylhet District, Bangladesh, where both the GB and the BRAC have been operating micro-loans over a period of several years.

Although social capital formation is an important dimension of the microcredit process, little attention has been given to the relationship of social capital to the microcredit pro-gramme in Bangladesh. Larance (1998) shed light on whether GB members' weekly meet-ings facilitate the members' ability to establish and strengthen networks outside their kinship groups and living quarters. She analysed network data and suggested that, by attend-ing weekly centre meetings, members can establish new networks and strengthen existing social ties that reach beyond their living quarters and familial group networks. Larance's study only focused on the relationship between women beyond their living quarters and kinship groups; it did not address the relationships between women within their own groups. Moreover, she did not explore whether the social collateral mechanism and peer monitoring of credit instalment payments at the due time increased conflicts and eroded exist-ing social capital among the impoverished borrowers. The aim of this study is to investigate whether women's participation in microcredit fosters relationships (i.e. social capital) between women within groups and beyond their groups in the rural area of Bangladesh. My interest lies in exploring whether microcredit membership and its collateral mechanism facilitate social networks, norms of reciprocity, collective identity and action among women borrowers in rural areas. Does the group principle work, or is it only a means for pro-curing instalments from the borrowers and a governmental strategy of microcredit organiz-ations, which follows the market rationality and recovers credit instalments at the due time?

Methods and data

The empirical data for this study have been collected from 151 married women microcredit borrowers of two microcredit project areas of GB and two project areas of BRAC in Sylhet,

Bangladesh, between June 2010 and December 2010. The project areas were purposively selected from three distinct villages: Zelegaon, Nodigaon and Shantigaon. I selected these villages as my research areas because both the GB and the BRAC have been operating credit programmes in these villages for about 4–18 years. I collected information only from those women borrowers who have been getting involved in microcredit programmes for one or more years from each project area.

Most of my data are qualitative and my analytical approach is descriptive, although I have also collected census data. I started my fieldwork by observation, by obtaining information from informal discussions with the respondents and also from their family members to familiarize with the villagers and identify the social issues in their communities. These discussions and observations helped in the development of a semi-structured questionnaire for collecting some quantitative/survey data. The survey covered the following aspects of the households of the respondents: sex and age composition, marital status, education and occupational structure, land ownership, duration of microcredit membership, current sums of credit and its investment.

In addition to the ethnographic observations and surveys, I also carried out ethnographic (in-depth) interviews and compiled case studies. When I became more familiar with each microcredit borrower's family I began to collect more qualitative data. First, I collected case reports from 35 village women borrowers who had been actively involved in microcredit programmes since their inception. These helped me to evaluate the population and build a mutual rapport with other microcredit borrowers' families, which aided me to proceed to ethnographic (in-depth) theme interviews. In theme interviews, I asked specific questions about social capital such as the factors that influenced the individual's participation in the credit programme, how women form their borrowing groups, whether they attend the group meetings, whether they know other group members and whether or not they get involved in mutual support and collective action. Each theme interview lasted for two hours or more. I collected information in Bengali, which is my native language. I regularly visited the weekly instalment collection places in the study area. Thus, the bulk of the data for this study was collected through in-depth interviews, case studies and observation. I use pseudonyms for the people I have quoted. I also avoid elaborating on the respondents' personal backgrounds and indicate which statements are from the same respondents.

Microcredit and social capital: problematic assumptions

Social capital is recognized as the missing link in development; a remedy to the social decay caused by neoliberal policies aimed at getting the prices right (Portes and Landolt 2000 in Mclean 2010). In the late 1990s, we witnessed a burgeoning of interest in, *inter alia*, social capital formation and discussions about applications of the concept across sectors and disciplines. However, the concept of social capital is neither an entirely new topic nor is it a monolithic concept. Rather, it has a long history in the social sciences particularly in sociology and political science. Scholars (Swedberg 1987; Portes and Sensenbrenner 1993; Portes 1998; Woolcock 1998) showed how the current theoretical foundations of the concept of social capital were derived from the writings of Alexis de Tocqueville, Durkhcim, Marx, Weber, Tonnies, Simmel and Talcott Parsons. French sociologist Pierre Bourdieu and American sociologist James Coleman established the systematic analysis and conceptual framework of social capital (Portes and Sensenbrenner 1993; Portes 1998; Portes and Landolt 2000; DeFellippis 2001). The concept of social capital first appeared in Bourdieu's work *La Reproduction* published in 1970. Bourdieu presented a systematic discussion of the concept in his writings under the title: 'The Forms of Capital' (1985). According to him, social capital incorporates obligation, the advantages of connections or social position and trust. Connections and obligations are not imposed, but are the product of investment strategies 'consciously or unconsciously aimed at establishing or reproducing social relationships that are directly usable in the short or long term'. Bourdieu also dealt with the interaction between capital, social capital and cultural capital and argued that all these were intertwined, convertible to economic capital, and that people purposefully invested in relationships that could bring them future benefits.

Bourdieu's clearest definition of social capital was considered as a theoretically useful and sophisticated attempt to deal with the issue, but the person who provided a comprehensive analysis of social capital in mainstream American sociology and brought social capital into widespread use in the social sciences is James Coleman (Portes 1998; Serageldin and Grootaert 2000; DeFellippis 2001). Coleman (1988) presented a sophisticated analysis of the role of social capital in the creation of human capital. Coleman defined social capital in terms of its function as a variety of entities each of which comprise some aspect of a social structure that in turn facilitate certain actions of individuals who are within the structure. He viewed social capital as productive and argued that, unlike other forms of capital (human capital and physical capital), social capital is inherent in the structure of relations between persons (Coleman 1990). For Coleman, social capital exists in the structure of relations among individuals and is thus largely intangible. Its potency can be realized in its capacity to facilitate individual action to a purposeful end. Social capital is different from human capital, which is embodied in individuals who possess knowledge and skills. As a result, they can perform a particular kind of action. Social capital is not embedded in the actors themselves; rather it exists in the structure of relations among actors (Coleman 1990).

Indeed, social scientists' explanations of social capital have focused on relationships between individuals or between an individual and a group. Such analyses emphasize individual development that flows from social capital, or how actors accrue benefit through their involvement in relationships or broader social structures (Bourdieu 1985; Coleman 1988, 1990). However, the attention of scholars later shifted to the role of social capital in community development (Portes 1998; Woolcock 1998; Portes and Landolt 2000). The principal source of the idea for community development practitioners and researchers is Putnam (1996, 1998, 2000). Putnam refers to social capital as 'the norms and networks of civil society that lubricate cooperative action among both citizens and their institutions' (1998, v). In Putnam's viewpoint, the central notion of social capital theory is that social networks have value: 'Social capital refers to connections among individuals, social networks, the norms of reciprocity and trustworthiness that arise from them' (2000, 19). Putnam also suggests that a society of many virtuous but isolated individuals is not necessarily rich in social capital. Therefore, voluntary associations are the 'features of social organization, such as networks, norms, and social trust, that enable participants to act together more efficiently pursue shared objectives' (Putnam 1996, 34). People's involvement in a community or association is an indispensable feature of social capital. Putnam argues that social capital is a resource that individuals or groups of people can either possess or not possess. He has argued that one of the important mechanisms for creating trust, norms, social engagement and cooperative behaviour is participation in networks of voluntary associations (Putnam 1993, 2000).

Since Robert Putnam's studies, a plethora of literature on social capital resulting in a myriad of conceptualizations has been published, but most of these studies are along the ideas of Bourdieu, Coleman and Putnam, which view social capital as located within networks, norms, trust and associational memberships (Ostrom 1994; Lin 1999; Serageldin and Grootaert 2000). In this paper, I have addressed the question as to whether women's associational memberships (i.e. microcredit memberships) solidify and foster social capital. By social capital I mean collective identity, actions and networks and how they facilitate norms of reciprocity and access to resources.

Formation of self-selected groups and access to credit

Microcredit programmes are based on the social collateral mechanism, in which one client stands as another's guarantor. In rhetoric, an individual poor woman cannot take a loan from a microcredit organization simply by asking for one. As a condition of receiving credit she has to be a member of a group (five to seven members form a group) and come to the centre (five to eight groups form a centre) for weekly instalments. The issue of group formation is very important theoretically in the sense that it is related to habitual social functions such as the facilitation of social networks, the development of a particular type of social norm of regularly attending weekly instalment meetings, making scheduled payments of credit instalments and democratically electing a group leader. Group formation can turn into a route for women microcredit borrowers to maintain regular interaction with other borrowers and the NGO's officials. Such interaction may nurture a sense of mutual understanding and help increase the chance of developing collective identity and action, thus allowing borrowers to establish and strengthen networks (both horizontal and vertical) and trust within and beyond their own homesteads or kinship groups.

The GB first started its activities in the village of Zelegaon about 18 years prior to the data collection for this study. The GB officials initially motivated families by saying that since they are poor they could take loans from their organization for running small business or other activities and pay back their loans through a number of weekly instalments but they would first have to organize at least 10 members to establish a group. Suchona (45 years old) had been serving as a GB group leader and also as the centre leader (*kendra prodhan*) of Zelegaon village since the inception of the centre. She informed me that when the bank first came to her area some people, including her husband, were motivated to take the loans provided by the GB official. Consequently, they took the initiative to develop the centre. She claimed that since all her neighbours were poor they also came forward to access the credit programme. In fact, when the centre was established in Suchona's house, she had also encouraged some of her neighbours to take out credit, otherwise the centre may not have been established here. I asked Suchona how she motivated others. Her response was:

> Actually, it was not a problem because they were not strangers. The group consisted of our neighbourhood (*para- protibeshy*) people; either they belonged to the same *gushti* (lineage) or they were related through kinship. First, we formed a group of 18 members, but after some months we took on four new members and excluded four old members from the group because they could not pay the instalments. They always caused trouble.
> [Q: Didn't you select the members carefully?]
> -Yes, when we formed the group we chose only those with whom we had good relationships (*moner mil ache*) and who would be capable of paying instalments. Yet, some members faced problems in paying back credit instalments.

Similar to that described by Suchona, the group and centre leaders of the BRAC and the GB of Shantigaon and Nodigaon villages confirmed that earlier, when microcredit started its operation, the NGO's staff encouraged some local people and then these people encouraged their neighbours to take credit. They unanimously stated that when they formed their borrowing groups they considered the members' sincerity and economic ability to pay instalments, the pattern of kinship networks and relationships. The respondents pointed out that they considered these characteristics to be important because if a member could not pay back instalments, who would take the responsibility? The self-selection practice underlying group formation led to the exclusion of the poorest of the poor and those who did not belong to the same *gushti* (lineage) or *bari*[2] (homestead).

It is no longer always necessary to motivate people to be members in the credit programmes because people now take credit on their own initiatives. The kinship networks and neighbourhood relationships are the key resources which, to a large extent, influence local people's participation in the microcredit programmes. I found that poverty, immediate economic constraints and natural calamities pushed rural people to be the members of microcredit programmes. The male members of the households usually send their wives, mothers, mother in-laws, sisters or other close relatives to access credit. The religious people (*mawlana* or *imam* in the mosque) used to avoid credit programmes because they had maintained that it is strictly prohibited in the Islamic religion to take or give loans with interest. However, even they send their female family members to take credit now. I found this to be a significant change in religious structure at a local level. However, when I sought answers to questions about credit from an *Imam* he avoided me, but the secretary of the mosque stated that they could not maintain this religious stricture because of poverty. However, among 151 married women microcredit borrowers, I found some members (20%) who informed me about the NGOs' field assistance and some old members of the NGOs had assisted them in joining the microcredit programme. I found that nowadays the NGO's officers or field assistants generally prefer to motivate two types of people to become members or maintain their existing membership – the economically well-off families who have a steady cash flow or income, and the members who have good payment records from earlier loans.

Today, the formation of a new group or self-selected solidarity group is no longer a requirement for taking credit from micro-lending organizations. Each eligible applicant who has sufficient individual collateral and has the capacity to pay back loans is only asked by the officials to collect two signatures from two female members and another signature from the centre chief to be accepted as a member. Excluding and including new members is a continuous process for the microcredit programmes and causes no anxiety among the new or old members since any sense of solidarity among the borrowers is low. The centre leaders play decisive roles in membership arrangements. They know the details about the inclusion and exclusion of members but the majority of the borrowers of a centre are not interested in who takes credit or not. The centre chiefs still keep in contact with many borrowers individually. Because of the collateral mechanism, the guarantors and the group leaders work to facilitate the programmes by collecting instalments from the defaulters and by disciplining the members. This sometimes causes scuffles and cultivates isolation and alienation among the members. As a result, leadership roles are generally unpopular among the rural women, who do not want the 'headache' of chasing up loans from their neighbours. The following statement from Nasibun's (a group and centre leader) exemplifies this:

> I think all the members do not know each other because some are not interested to know who takes credit or who quits the programme. But I have to keep them in mind, because I have to ensure their repayments every week. Sometimes it becomes a great hassle for me. It breeds conflict with others. If any borrower fails to pay the instalments I have to talk with her, motivate her to pay back credit, which is a waste of time for me. I have decided to cut my name from the programme, so I get a relief from this burden.

Group meeting norms and the myth of collective identity

Group meetings or weekly credit instalment meetings are considered to be the main route to building particular social norms and social capital among the microcredit borrowers. Scholars (Ostrom 1994; Schuler and Hashemi 1994; Jain 1996; Bastelaer 2000; Dowla 2006) have argued that social capital or a particular cultural habit is facilitated when microfinance organizations, such as the GB and its replicators, require all members to follow the bank norms as a matter of routine every week. Such rituals include saluting and reciting the list of decisions that accompany group membership. When microcredit programmes started up in this study's research, areas within the GB in particular emphasized the importance of group meetings or borrowers' presence in weekly instalment payment places. However, today, this requirement has completely lapsed. The organization no longer gives any importance to group meetings. The role of these meetings is now limited to providing credit and recovering instalments. Suchona (aged 45), a GB microcredit borrower and the group leader of the village Zelegaon, has keen observations on the transition of the GB micro-lending business in her own locality: 'Now in Zelegaon a number of banks (microcredit organizations) provide credit, so GB has stopped caring about rules and regulation because the borrower could go to another NGO and as a result, the GB would lose its clients.'

Like Suchona, most borrowers repeatedly report that both GB and BRAC are now reluctant to follow their own rules and regulations. All the respondents of their respective villages unanimously stated that they had not seen their organizations calling any meetings (*karmoshala*) at the centres for several years. For example, Helena (a microcredit borrower of Zelegaon village) recounted that when she first took a loan from GB about 18 years ago, the officer told her to read the 16 decisions of GB to understand the rules, but now GB officials do not say anything about the 16 decisions. For many years, Helena did not witness the GB call any meetings (*karmoshala)* or offer any training programmes for its borrowers. The officials only advised members to take more credit and update credit instalments (*shomoy moto ksiti dao)*. Helena observed that, more recently, some borrowers took a substantial amount of credit for buying a three wheeler but in practice they worked as money lenders. According to her,

> What do we get from this GB business? Nothing: does GB feel our pains? Every week the GB officer comes to my adjacent for collecting weekly instalments but never asks me what I ate today? How can I pay the instalments? How do I spend my days? What is the future of my children? Why could not our children go to school? Or why do I give dowry for marriage?

I have observed that many women send their credit payments through their husbands or close relatives. Some women never visit the payment centres. They only authorize their husbands or sons to manage the formalities in the centre. The economically well-off families and some older men never send their wives to pay the weekly instalments. Often, borrowers' husbands visit the centres to pay the instalments and in some cases the microcredit officers themselves collect the credit instalments from the borrower's place of business. Consequently, I argue that microcredit programmes in the present study's research area fail to facilitate group meeting norms, fail to develop social conscientization activities and do not promote a sense of solidarity among the borrowers.

Minara (aged 45) explained that members actually send their credit instalments when it suits them, depending on their work schedules. According to her, the officer collects weekly instalments (*kisti*) in her centre generally between 9 and 11 am. Consequently, most borrowers do not visit the centre at the same time. Moreover, Minara thinks that since some women do not prefer to come to the centre they are not aware of who takes credit and who has left the programme. Minara does not attend loan collecting meetings every week. I enquired as to why this was so, and she replied:

> What is the necessity to go there? I only know that I have to give the instalments money every week, nothing more. If you have nothing nobody will help you. As I have taken out credit for several years, I have been continually interacting with some borrowers. Despite the fact that most of them are my neighbours; I only socialize (*chola-pera*) with them.

In contrast to the above encounters, I found that there are some women who do regularly visit the centres for credit instalment payments. Those who do attend say that they feel good about the opportunity to talk with other women and to inquire about other people's news. They share ideas and develop acquaintances with women outside their kinship groups or hamlets. At least once in a week they can wear their finest dresses and small jewellery ornaments. They are also starting to adopt new styles. For example, some women have followed one another in using lipstick. As one microcredit borrower says:

> When some women walk the road side to go and give their instalments (*kisti*) wearing their good *saris* it looks really beautiful. In the centre, they can talk to each other, which helps to develop face to face relations (*chokha chokhi shomporka*).

However, whenever I asked whether such relationships could bring any mutual productive benefit or help to form collective identity I was amazed as to how pessimistic their outlooks were. Most of the women expressed the view that credit relationships cannot bring any mutual or shared benefit, or engender any collective identity or norms of reciprocity. For example, Jesmin (39 years old) admitted that she knew most of the group members because she got involved in microcredit when the BRAC started its operation in her area. She actively worked to develop this centre and has tried to maintain good relations with other members, but she feels isolated:

> Here all belong to the same *gushti* (lineage) that is, *maimol* (fishery). Some of them have good relations and unity but I am alone, I am *abadi* (an outsider). So here people even many group members do not like me, they are envious (*hingsha kore*) of me. Sometimes they quarrel with my family members.

Jesmin said that her family originated from Jessore. They bought homestead land in the village and built a house many years ago, but the inhabitants often referred to her family as *abadi*. Jesmin's husband is a member of the village *panchayat*, but the neighbours and some borrowing group members do not like her family.

When I asked my respondents whether or not they think that the members of their group are equal, they responded that they are equal as human beings, but in terms of mentality and to some extent socio-economics, they are not. 'Not everyone can *samaj* (socialize or make associations with others) with everyone else', they said. They often asked me if the five fingers of a hand are equal. Most of my respondents argued that the poor people always have less unity and solidarity towards their neighbours, and a poor person cannot tolerate another poor person's good situation or happiness. They said that if a poor family can improve their economic condition, then they can soon forget their poor kin, as also noted by Streefland (1986). They also noted that when a rural poor family gets rich, other families become envious (*hingsha*) of their wealth and try to bring them down (*tene niche namate chai*).

I will next explore whether microcredit-providing NGOs have neglected or nurtured democratic practices and promoted leadership among the women at the centres. Larance (1998) found that the centre chief changed each year, but in the research area I studied, I found that the same women had been serving as the centre chiefs/group leaders for about 4–18 years. For example, Suchona had been serving as the centre chief of a GB microcredit centre for the last 18 years. Shorola had worked at the GB centre in Shantigaon village for about nine years. Likewise, Zarina and Alapi had been serving as the centres' leaders for the last four years of the BRAC microcredit programme. Changes in leadership were rare and leaders were not elected democratically. The micro-lending officials usually select leaders who can give a significant amount of time to the NGO work, including collecting weekly instalments and overall governance. A centre leader must be clever, vocal and have the capacity to pay back her own instalments when they are due.

Collateral mechanism and market rationality

Microcredit-providing NGOs operate according to market logic. I asked my respondents why NGOs disbursed credit through the collateral mechanism wherein a new applicant has to collect the signatures of two other women borrowers and also from the group leader as guarantors. Most of my respondents noted that micro-lending organizations must collect signatures from two women in addition to a signature from the group leader in order to run their credit businesses smoothly in the local communities. According to these respondents, the NGO officials are not their fellow villagers, so they do not know all the villagers. When an individual applies for a loan they cannot sanction the loan on behalf of unknown people. In order to recover instalments at the due times without facing any obstructions, they disburse credit through guarantors and the group leader. Some of my respondents declared that when a member defaults, the NGO officials use a group of women to recover instalments, which causes conflicts, alienation and destroys prevailing social relationships. The respondents often stated that it is their (NGO's) goal to fill their own bellies with the help of others. On the basis of my ethnographic data, I argue that the mere formation of borrower groups among women serves only as a tool for collecting credit instalments and not for strengthening ties between the members. The following statement from Afia (aged 36) exemplifies women's views:

> It (collateral mechanism) is a process of frying fish in its own oil. When anyone wants to take credit, the NGO's officer compels her to collect two other women's signatures and the signature of the group leader. As a result, three persons including their own family members are involved in one woman's loan. Now if a woman fails to pay instalments at least five people go at a time to press that sole woman to pay the instalment. That creates brawling, conflict and the need for arbitration among the members. The NGO officer does not have to face any problems and spend extra time in collecting the instalments.

The above statement suggests that microcredit programmes engage in a group mechanism only in the most instrumental manner and as a governmental strategy to reduce administrative costs and to maintain financial discipline. The group mechanism is used for collecting instalments and to improve the micro-lenders financial health rather than the welfare of the borrowers, as also mentioned by Rankin (2002) in a study on microfinance and social capital in Nepal. Like other scholars (Schuler and Hashemi 1994; Larance 1998), I do not see microcredit programmes generating a sense of mutual interdependence, confidence and willingness to help each other. I found no clear example of joint-liability arrangements by which group members took responsibility for other member's repayment problems in the researched areas. Defaulters generally rely on kin and close friends rather than on fellow members to pay instalments. Most members do not consider their guarantors or groups as mutual helpers; instead, they view them as pressure groups. Some other researchers also reported that 'staff pressure' sometimes led to violent collective action by fellow members against defaulters (Montgomery 1996; Ito 2003).

I assumed that Fuglesang and Chandler's (1993, 100) description of the GB's activities could be the normal set-up everywhere in Bangladesh: '[… Field staff] use every occasion to reinforce the message: You must go forward together and help each other'. However, I found that the microcredit organizations in my research areas operated along the principle of the neoliberal ideology. An individual borrower is responsible for her own success or failure of credit investment. The officials do not give any advice about the utilization of credit. The NGO's field staff get involved in three main duties: disbursing credit, bringing in new members to a group and recovering instalments at the due dates.

What struck me is that many borrowers viewed the microcredit organizations first and foremost as money-lending businesses (*shud-babshaw*). They referred to microcredit organizations as the instalment collecting (*kisti toler*) bank, and the weekly meeting place as the loans or instalments collecting centre (*kisti toler Kendra*). All the borrowers unanimously considered their debts (*riin*) as private. A member must solve her own repayment problems by herself. On the basis of memberships and economic ability, I defined three categories of microcredit borrowers in the research areas of this study. The first category (25%) consists of a small segment of borrowers who are economically well off (such as school teachers, rural doctors, household members abroad and those who have lucrative businesses or own a sufficient amount of land), have maintained their memberships for many years and paid instalments smoothly. The second category consists of a large number of microcredit borrowers (65%) who are economically poor and have been suffering from a vicious circle of indebtedness over the years. They maintain their regular payment schedules or often pay off previous loans by taking out new ones. The third category (10%) includes the economically well off and some poor families who generally do not continuously maintain their memberships, but maintain their savings account because it works as a security for taking credit when they might need it in the future.

Each individual is responsible for paying off credit and there is no means to escape from credit repayments. The NGO's field assistants have extensive networks and door-to-door contact for collecting instalments. I observed a new kind of patron–client relationship between loan officials and borrowers. The nature of the relationship between the microcredit borrowers and the officers is hierarchical and built on an unequal division of power, for instance that of the creditors and the powerful and that of the weak debtors. Since the solidarity and collective identity among the borrowers is lacking, the patron (NGO officials) can use his discretionary power to levy fines for defaulting, apply sanctions or deny the biggest window of capital (credit) to the poor at will. These activities have also been explored by Ito (2003) in her research on the GB in Bangladesh.

Some of my respondents told me that it is counterproductive to query members' credit investment practices as such enquiries can lead to tension. Ackerly (1997) has also reached the same conclusion elsewhere: when group members do monitor one another's consumption and repayment patterns in accordance with programme incentive structures, this can lead to hostility and conflict instead of unity. I found that there was a backstage negotiation between group leaders and some members about the using of credit. Some borrowers took a substantial amount of credit for buying a three wheeler, but in practice they handed loans to other relatives at high interest.

Solidarity is supposed to provide a productive or valued resource flow. Hence, it is what sociologists have characterized as social capital (Coleman 1990). In the same vein, Putnam (1993, 2000) has also argued that membership of an association reinforces the importance of cooperative behaviour and attitudes for the mutual benefit of the group's participants. However, it is not obvious as to which situations members of a group become involved in reciprocal relations. The overwhelming majority of the respondents in this study declared that they do not have the economic ability for reciprocity. Because of their appalling economic situations they cannot get involved in such relationships. I found that altruistic help (*shahaja*) among the borrowers is rare. The following statement by Nasibun (aged 38) exemplifies women's views:

> The members have no helping attitude. Nobody wants to help others without interest. If any member helps you, s/he is always awaiting when you also help her in return. If you fail to offer similar help, then the relationship gets destroyed. If anybody has money, s/he also does not wish to give a loan, rather he /she gives advice to take a loan from *kisti* (instalment) bank. This is the rule (*niyom*) of the world.

To a certain extent, the neighbours and relatives may help each other out – for example, providing a pot of rice or kitchen equipment. However, their economic situation does not permit them to help others altruistically. The analysis of data indicates that microcredit has brought a change in the local web of ties in the poor localities. The people who have economic ability do not lend money due to the fear of not getting back the money at the due time; instead, they suggest that the others take loans from microcredit organizations. As a result, the relationship gets destroyed and people get involved in the microcredit programmes.

The analysis of this study data indicates that poverty and competition over limited resources erodes the fabric of social life and solidarity in rural Bangladesh. Although personhood is entangled in familial and kinship relationships, brawling, disunity, enviousness, expectations, competition and individualistic self-interest are an inherent part of normal familial life for the poor in Bangladesh (Streefland 1986). As in Larance's (1998) observation, I did not find that membership in the credit programme mobilized the community and created a platform or collective identity among the members. I argue that although microcredit programmes depend on existing social capital, they fail to create more social capital.

Conclusion

This article contributes to the understanding of whether microcredit and its collateral mechanism facilitate social networks, enhance norms of reciprocity, improve collective identity and increase action among the women borrowers in the rural areas of Bangladesh. As mentioned before, the social capital school (Putnam, 1993, 2000) has proposed that one of the important mechanisms for the generation of trust, norms of reciprocity, civic engagement and cooperative behaviour is participation in networks of associations and organizations. The findings of the present study indicate that microcredit membership in the area that I have studied fails to promote women's social capital or community mobilization. These findings are despite the fact that both the GB and the BRAC depend on prior personal relations among their members as the basis for group formation. Although social capital plays a decisive role for the formation of borrowing groups and access to credit, the analysis of data suggests that the social collectiveness, mutual support and economic exchange among the borrowers are, overall, very scarce. Group membership, joint liability, shared objectives or the attitude towards the importance of group can hardly be regarded as expressions of social connectedness.

The social collateral mechanism, by and large, works as an entry route to access a credit programme for the borrowers, but for NGOs it is an instrument for promoting their governance and disciplinary power over the borrowers and furtherance of capitalist gains, such as collecting instalments at the due times. Meeting to discuss problems or joining weekly instalments meetings are no longer part of the practices of microcredit organizations. Putnam (1993) notes that networks of civic engagement are essential forms of social capital – the denser such networks in a community, the more likely its citizens will be able to cooperate for their mutual benefit. To my mind, 'quality and purpose' (Kabeer, Mahmud, and Castro 2012) of networks are much more important than the 'density of networks'. The borrowers refer to microcredit organizations as the money-lending business (*shud-babshaw*) or instalment collecting (*kisti toler)* banks and weekly meeting places as loans or instalments collecting centres (*kisti toler kendra*). An applicant's ability (individual collateral) to repay her loan is the main criteria for her access and acceptance into a credit programme.

The package of training and consciousness-raising lessons that originally went alongside the GB microcredit programme is now missing from their rural operations. Microcredit organizations have turned into commercial/business-oriented organizations that follow the self-sustainability paradigm. Thus, they have narrowed their activities to the sanctioning of loans and collecting instalments on those loans. As a result, they have failed to achieve the following: to develop social conscientization activities, to build a vision of collectivities of working together or to reinforce norms of solidarity among the women borrowers. Microcredit programmes have not strengthened solidarity or favoured the creation of solidarity in the poor localities.

Poverty, scarcity of work and competition over limited resources among the impoverished microcredit borrowers are inherent parts of normal family life and scarcity has worsened (Streefland 1986; Jansen 1987). These problems are likely to grow and generate an environment of envy and hostility, which will fragment social relations rather than bring people together. Microcredit programmes tend to reinforce and even intensify existing socio-economic inequalities. Neither microcredit programmes nor the government of Bangladesh offer any intermediary programmes or financial intervention through which the existing inequalities could be lessened. Therefore, I argue that the assumed association between microcredit group membership and building social capital (new social networks, norms of reciprocity, and collective identity and action among microcredit borrowers) is much less prominent than is commonly assumed by many scholars and development practitioners.

Acknowledgements

I am grateful to Sirpa Tenhunen for her encouraging, skilful supervision and constructive comments on the PhD dissertation from which this article is drawn. I also thank both Sirpa Tenhunen and Minna Säävälä for their helpful comments on the first draft of this article. This study was made possible by research funding from the Academy of Finland [grant no. SA 138232] and a scholarship from the Jenny and Antti Wihuri Foundation. I would like to thank the funding agencies for their support.

Notes

1. The term 'group lending' generally means a process whereby individual loans are given to a small, self-selected and homogenous group of borrowers (five to seven members) who are collectively responsible to loan repayment.

2. A *bari* is a collection of households which are divided and sub-divided, and they are almost always patrilineally linked (Gardner 1992; Todd 1996). *Bari* clusters together form a *para* or neighbourhood (Todd 1996).

References

Ackerly, Brooke. 1997. "What's in a Design? The Effects of NGO Programme Delivery Choices on Women's Empowerment in Bangladesh." In *Getting the Institutions Right for Women in Development*, edited by Anne Marie Goetz, 140–160. New York: Zed Books.

Adams, D. W., D. H. Graham, and J. D. von Pischke, eds. 1984. *Undermining Rural Development with Cheap Credit*. Boulder, CO: Westview Press.

Adams, D., and R. Vogel. 1986. "Rural Financial Markets in Low-Income Countries: Recent Controversies and Lessons." *World Development* 14 (4): 477–488.

Anderson, C., Laura Locker Leigh, and R. Nugent. 2002. "Microcredit, Social Capital, and Common Pool Resources." *World Development* 30 (1): 95–100.

Bastelaer, T. Van. 2000. *Imperfect Information, Social Capital, and the Poor's Access to Credit*. IRIS Center Working Paper No. 234, University of Maryland, Center on Institutional Reform and the Informal Sector (IRIS).

Binswanger, Hans P., and P. Landell-Mills. 1995. *The World Bank's Strategy for Reducing Poverty and Hunger: A Report to the Development Community*. Environmentally Sustainable Development Studies Monograph 4. Washington, DC: World Bank.

Bourdieu, P. 1985. "The Forms of Capital." In *Handbook of Theory and Research for the Sociology of Education*, edited by J. G. Richardson, 241–258. New York: Greenwood.

Coleman, J. S. 1988. "Social Capital in the Creation of Human Capital." *American Journal of Sociology* 94: S95–S120.

Coleman, J. S. 1990. *Foundations of Social Theory*. Cambridge, MA: The Belknap Press of Harvard University Press.

DeFellippis, James. 2001. "The Myth of Social Capital in Community Development." *Housing Policy Debate* 12 (4): 781–806.

Dowla, A. 2006. "In Credit We Trust: Building Social Capital by Grameen Bank in Bangladesh." *The Journal of Socio-Economics* 35 (1): 102–22.

Fuglesang, A., and D. Chandler. 1993. *Participation as a Process – Process as Growth: What We Can Learn from Grameen Bank of Bangladesh*. Dhaka: Grameen Trust.

Gardner, K. 1992. "Migration and the Rural Context in Sylhet." *New Community* 18 (4): 579–590.

Hoff, Karla, and Joseph E. Stiglitz. 1990. "Introduction: Imperfect Information and Rural Credit Market: Puzzles and Policy Perspectives." *The World Bank Economic Review* 4 (3): 235–251.

Holt, Sharon L., and Helena Ribe. 1991. *Developing Financial Institutions for the Poor and Reducing Barriers to Access for Women*. World Bank Discussion Papers 117. Washington, DC.

Ito, S. 2003. "Microfinance and Social Capital: Does Social Capital Help Create Good Practice?" *Development in Practice* 13 (4): 322–332.

Jain, P. S. 1996. "Managing Credit for the Rural Poor: Lessons from the Grameen Bank." *World Development* 24 (1): 79–89.

Jansen, E. 1987. *Rural Bangladesh: Competition for Scarce Resources*. Dhaka: The University Press.

Kabeer, N., S. Mahmud, and J. G. I. Castro. 2012. "NGOs and the Political Empowerment of Poor People in Rural Bangladesh: Cultivating the Habits of Democracy?" *World Development* 40 (10): 2044–2062.

Larance, L. Y. 1998. *Building Social Capital from the Centre: A Village-Level Investigation of Bangladesh's Grameen Bank*. Centre for Social Development: Working Paper No. 98 (4). St. Louis, MO: Washington University in St. Louis.

Lin, Nan. 1999. "Building a Network Theory of Social Capital." Paper presented in International Sunbelt Social Network Conference, Charleston, SC, February 18–21.

Mayoux, L. 2001. "Micro-finance for Women's Empowerment: A Participatory Learning, Management and Action Approach." Paper presented at the Asia Regional Micro-credit Summit, Milton Keynes, UK, UNIFEM, February.

Mayoux, L. 2002. "Microfinance and Women's Empowerment: Rethinking 'Best Practice.'" *Development Bulletin* 57: 76–80.

McGregor, J. A. 1988. "Credit and the Rural Poor: The Changing Policy Environment in Bangladesh." *Public Administration and Development* 8 (4): 467–482.

Mclean, K. 2010. "Capitalizing on Women's Social Capital? Women-Targeted Microfinance in Bolivia." *Development and Change* 41 (3): 495–515.

Montgomery, R. 1996. "Disciplining or Protecting the Poor? Avoiding Social Costs of Peer Pressure in Micro-credit Schemes." *Journal of International Development* 8 (20): 289–305.

Ostrom, E. 1994. "Constructing Social Capital and Collective Action." *Journal of Theoretical Politics* 6 (4): 527–562.

Portes, A. 1998. "Social Capital: It's Origins and Applications in Modern Sociology." *Annual Review of Sociology* 24: 1–24.

Portes, A., and P. Landolt. 2000. "Social Capital: Promise and Pitfalls of Its Development." *Journal of Latin American Studies* 32 (2): 529–547.

Portes, A., and J. Sensenbrenner. 1993. "Embeddedness and immigration: Notes on the Social Determinants of Economic Action." *American Journal of Sociology* 98 (6): 1320–1350.

Putnam, R. 1993. *Making Democracy Work: Civic Traditions in Modern Italy.* Princeton, NJ: Princeton University Press.

Putnam, R. 1996. "The Strange Disappearance of Civic America." *American Prospect* 1996: 34–48.

Putnam, R. 1998. "Foreword." *Housing Policy Debate* 9: v–viii.

Putnam, R. 2000. *Bowling Alone: The Collapse and Revival of American Community.* New York: Touchstone.

Rankin, Katharine N. 2002. "Social Capital, Microfinance, and the Politics of Development." *Feminist Economics* 8 (1): 1–24.

Schuler, Sidney Ruth, and S. M. Hashemi. 1994. "Credit Programs, Women's Empowerment, and Contraceptive Use in Rural Bangladesh." *Studies in Family Planning* 25 (2): 65–76.

Serageldin, I., and C. Grootaert. 2000. "Defining Social Capital: An Integrating View." In *Social Capital: A Multifaceted Perspective*, edited by P. Dasgupta and I. Serageldin, 40–58. Washington, DC: World Bank.

Sobhan, R. 1998. *How Bad Governance Impedes Poverty Alleviation in Bangladesh.* Working Paper No. 143. OECD Development Centre.

Stiglitz, Joseph E. 1990. "Peer Monitoring and Credit Markets." *The World Bank Economic Review* 4 (3): 351–366.

Streefland, Pieter. 1986. *Different Ways to Support the Rural Poor: Effects of Two Development Approaches in Bangladesh.* Dhaka: The Centre for Social Studies, University of Dhaka.

Swedberg, R. 1987. "Economic Sociology: Past and Present." *Current Sociology* 35 (1): 1–25.

Todd, H. 1996. *Women at the Centre: Grameen Bank Borrowers after One Decade.* Dhaka: The University Press.

Woodworth, W. P. 2008. "Reciprocal Dynamics: Social Capital and Microcredit." *ESR Review* 10 (2): 36–42.

Woolcock, M. 1998. "Social Capital and Economic Development: Towards a Theoretical Synthesis and Policy Framework." *Theory and Society* 27 (2): 151–208.

Yaron, J. 1994. "What Makes Rural Finance Institutions Successful?" *The World Bank Research Observer* 9 (1): 49–70.

Mobile telephony, mediation, and gender in rural India

Sirpa Tenhunen[a,b]

[a]Nordic Centre in India, Communication and International Relations Office, University Liaison Building, Umeå University, Umeå, Sweden; [b]Department of History and Ethnology, University of Jyväskylä, Jyväskylä, Finland

This article aims to develop the understanding of new media and social change by examining how mobile phones mediate kinship and gender in rural India. I provide a nuanced picture of the contested nature of kinship and gender in the village based on long-term fieldwork in order to explore how mobile phones mediate relationships and ongoing processes of social change. The article illustrates how the physical qualities of phones help strengthen the multiplicity of discourses by mediating relationships and contributing to the multiplicity of speech contexts. Mobile phone use has been encouraged and motivated by kinship relationships and the use of mobile phones has, in turn, transformed these relationships by helping to create new contexts for speech and action. However, instead of the drastic improvements or changes, for instance in economic power relationships, the positive impacts of women's phone use appear subtle and ambiguous: most calls are about the slight redefinition of the home boundaries.

This article aims to develop the understanding of new media and social change by examining how mobile phones mediate kinship and gender in rural India. Mobile phone users in developing countries have emerged as iconic figures signifying change and progress as mobile telephony has triumphed unexpectedly rapidly. Ninety percent of the world is currently covered by mobile networks and the ownership of mobile phones practically tripled in developing countries between 2002 and 2006 (International Telecommunication Union 2010). The growth of India's mobile phone density has been among the fastest: teledensity increased in India from less than 1 per 100 persons to 78 per during 1991–2012 (Telecom Regulatory Authority of India 2012). Few studies (Tenhunen 2008a; Doron 2012; Tacchi Kathi, and Crawford 2012; Jouhki 2013) have shown that women in different parts of India have experienced the opportunity to use mobile phones as a major asset although, in comparison to men, their access to mobile telephony tends to be more restricted. I aim to provide a nuanced picture of the contested nature of kinship and gender in the village based on long-term fieldwork in order to explore how mobile phones mediate relationships and ongoing processes of social change.

That women's use of mobile phones can be a contested issue in India became headline news worldwide when international news agencies reported that the Sunderbari village council of the state of Bihar banned women from using mobile phones. The village elders felt that phones polluted the social atmosphere and enabled elopements (Burke and Kumar 2012). Kärki (2013) observed similar reactions to women's phone use in rural Rajasthan. Parents there explained that they wanted to have their daughters married off under age to prevent them from talking over the phone with boys and eloping. People in rural Rajasthan told Grodzins Gold (2009) that phones spoil relationships as face-to-face meetings are being replaced by phone calls. Doron (2012) notes the destabilizing nature of mobile phones for social relationships in Banares maintaining that phones can be incorporated into households in ways that reaffirm dominant norms and practices by restricting young women's phone use. In his study on rural Tamil Nadu, Jouhki (2013) found men to be more active, dominant, and technologically literate users of mobile telephony than women. He argues that young men enjoyed more freedom and agency in the sphere of mobile telephony than women.

In common with Jouhki (2013) and Doron (2012), much of the anthropological research on the uses of new technologies share an emphasis on technologies tending to reinforce existing structures and, especially, adherence to kinship patterns (Horst and Miller 2006; Barendregt 2008; Archambault 2010). The few studies on the women's mobile phone use in the global South indicate both potential for change, as well as the perseverance of gendered patterns. Chib and Hsueh-Hua Chen (2011) demonstrate how female mobile phone users in Indonesia maneuver through their social constraints to reap benefits from information and communication technology (ICT) use. Wallis (2011) argues that the women's gender, age, class, and rural origin produce particular constraints on their ability to generate higher income and find better jobs in Beijing. By studying the positionality of users in various contexts and the way that socio-techno practices arise within prior social worlds, she sheds light on why marginalized workers' deployment of mobile phones will not necessarily lead to greater income, a better job, or more autonomy.

Gender and technology studies have shared an interest with anthropological studies on mobile communication in seeking to understand why and how modern Western technology has emerged as a male domain (Oakley 1974; Cockburn 1983; Wajcman 1991). As Wajcman (1991) points out, feminists have long viewed the symbolic representation of technology as sharply gendered. Commanding new technologies is a highly mythologized and valued activity since it signifies being involved in directing the future. Consequently, men's affinity with technology is integral to the constitution of male gender identity and the culture of technology. As seen by the constructivist framework, the relationship between technology and society, as well as between gender and technology is mutually constituted: the central tenet of feminist technology studies is that specific technological artifacts may be gender shaped and may have gender consequences. The process can be explored by examining the design and use of technologies (Lohan and Faulkner 2004). Women's everyday uses of technology can, in turn, relate to how gender is constructed.

Research on the social uses of the telephone in western countries has systematically shown clear gender differences: women use the telephone at home more often than men (Moyal 1992; Ling 1998).Yet, at times, supposedly masculine technologies have been appropriated for distinctively feminine ends, as Fischer's (1992) social history of the telephone in North America up to World War II illustrates. In comparison to anthropological studies on ICTs, it is the studies on reproductive technologies which have addressed the disruptive effects of technology. Strathern (1992) broke new ground in anthropology by arguing that new reproductive technologies could destabilize ideas of nature.

Drawing from Schneider (1980), I perceive kinship as a symbolic system consisting of indigenous categories of social organization but in a post-Schneiderian vein. Following Collier and Yanagisako (1987), I explore gender and kinship as interrelated domains which enable fluidity and agency and relate to power issues. I am interested in how mobile telephony is influencing and drawing from local social and cultural as cultural practices. I follow Schatzki, Knorr-Cetina, and Savigny (2001), Reckwitz (2002), and Shove, Pantzar, and Watson (2012) in viewing material objects as key part of practices and I analyze the role of materiality through the concept of mediation. I also go beyond anthropological works on practice, such as those by Bourdieu (1992), Sahlins (1987), and Ortner (1989), by paying attention to actors' critical faculties, disjunctures, and discontinuities in daily practices. I explore the gendered use of technology as part of the co-production of technology and gender: how is the use of mobile phones gendered and how do mobile phone-mediated conversations and interactions feed into gendered processes of social change?

Defined broadly, the concept of mediation refers to how a given medium reconciles the various forces of history, culture, and the material world and both constrains and enables social actors' use of that medium. This flexible notion has recently been used in many ways to allow us to understand the role of new media in communication and social interaction, as well as in social and cultural changes related to the intensification of media. The term has been used interchangeably with mediatization (Altheid and Snow 1988, 195), whereas others have reserved the term mediation to refer to the role of mass media in transforming society (Schulz 2004, 89). Hjarvard (2008a, 2008b, 13–14) restricts the use of the term mediatization to recent phenomenon, distinguishing it from mediation, by which he means the use of any medium to achieve communication. However, many scholars currently understand mediation as communication through media (Encheva, Driessens, and Verstraeten 2013). In this article, I build on the view endorsed by anthropologists that mediation need not be assigned only to media technologies because it can be regarded as a general condition of social life (Mazzarella 2004; Boellstorff 2008; Horst and Miller 2012). I view all interactions as mediated in the sense that interaction and speech are always influenced by their contexts. As Horst and Miller (2012) argue, there is no pure human immediacy, as all interaction is as culturally inflected as digitally mediated communication. People commonly respond effortlessly to changes in contexts: they have fairly clear ideas of what can be expressed and how in whose presence. The question is not how unmediated culture becomes mediated through new media, but how different forms of mediations interact when a powerful new medium is appropriated. Following Latour (1999), I acknowledge that the materiality of mobile technology plays a role in facilitating and mediating certain types of interaction and activities.

The article is based on interviews, observation, and survey data on the use of mobile phones in rural West Bengal in 2005, 2007–2008, 2010, and 2012–2013. During these fieldwork visits, which ranged from 2 to 4 months (altogether 11 months), I observed phone calls as part of daily activities mainly in one neighborhood in the village of Janta. Observation has been a key method as it would not have been possible to study especially subversive phone uses through interviews and my study complements those studies on the gendered use of mobile phones that mainly draw from interviews with phone users (Doron 2012; Jouhki 2013). As male researchers Doron (2012) and Jouhki (2013) have gained more information about men's calls than what was possible for me as a woman in a village setting. Perhaps as a woman I was in better position to observe women's phone use than a male researcher could have been and this article focuses more on women's than men's phone use although I regard gender as a relational concept. I maintain that even if women use mobile phones less than men and mainly call their relatives, it is worth observing what kind of conversations women carry out by phone because women find these calls significant and there is a chance that their calls are part of discursive change which may eventually lead to epochal changes.

However, I also find it important to form a broader picture of calling patterns through interviews. My quantitative research material consists of interviews with phone owners (130 personally conducted, free-flowing interviews and 121 structured interviews carried out by a research assistant in a few adjacent villages[1]) in 2007–2013, the filming of 100 phone calls from public phones in Janta, the phone diaries of 27 families in the Tili neighborhood that a research assistant Rekha Kundu collected in 2011, and a survey of 158 households in Janta in 2010.[2] I start by relating phone use to other communication practices use, as well as to kinship and gender relationships in the village, which represent interrelated contexts of mediation. Next, I describe the gendered diffusion of phones and phone use patterns. Thereafter I analyze how phone use draws on and contributes to the ongoing changes by mediating speech and connecting different spheres of mediation.

Communication before phones

Janta is a multi-caste village with 2328 inhabitants (author's census 2004) in the eastern state of West Bengal in India. The Tilis (50%) are the dominant caste, both numerically and in terms of land ownership. Other major caste groups are the Bagdis (15%) and Casas (16%). Most Tilis and Casas own land, while most Bagdis, who are classified as a scheduled caste, earn their livelihood by means of daily labor: agricultural work or work in the brick factories. Unlike its neighboring states, West Bengal is not among the poorest states of India. During 1970–2000, poverty reduction in West Bengal was among the fastest in India (Planning Commission 2012), but it has not fared as well in comparison to other parts of India in terms of infrastructure or providing education and health care. The greatest achievement of the communist rule in West Bengal (1977–2011) was the land reform. Although the plots of land distributed to the landless have been too small to sustain entire families throughout the year, land reform has provided landless laborers with additional means of livelihood.

The social networks of rural West Bengal, India's most densely populated state, have always been well knit. Local identity has emerged in relation to broader networks – other villages, districts, and cities – through trade and kinship. Connections did not start with the use of information technology, although phones, as well as radios and television sets have increased and intensified such links. Village exogamy is responsible for most of the connections between villages. My sample of 67 women, of whom 66 had married outside their natal village, illustrates that village exogamy is the dominant marriage form in the region (Tenhunen 2008a). Village exogamy leads to connections between villages since marriages are not an end in themselves but a form of alliance creation – post-marital visitations and prestations are the essence of marriage alliances (Fruzzetti 1990, 37). Women do not cut their ties with their natal families when they get married, as the two lines continue to interact in ritual, as well as in other contexts.

Before the arrival of phones, news of relatives in other villages, of women's natal families and men's in-laws, was conveyed via letters (by the literate part of the village) and visitors. Visiting also meant being informed of other people's news (*khobor neua*) and delivering it to the relevant people in other villages, often on request. However, these visitor networks were often slow and even unreliable. Before the phone system was established, villages were relatively isolated in comparison with urban areas. Especially women stressed that the main benefit of phones for them is that they could not previously obtain news of a serious illness or the death of a close relative in time to view the body before the cremation and to participate in the death rituals. The inadequate communication infrastructure contributed to villagers' considerable autonomy from state authorities and services. The more remote the village, the smaller the chances were of asking for or getting help from outside in the case of emergencies and conflicts.[3]

Gendered diffusion of phones

It was a telling sign of the potential of phones for women that the first phone in the village was acquired by an elderly woman's sons for her. The sons had emigrated to work outside the village and the phone helped them stay in touch with their mother. It is, however, equally revealing that the first group of mobile phone buyers were men: micro entrepreneurs, car, and tractor drivers who found phones useful for staying in touch with customers and calling for help if they experienced problems on the road. By 2007, the number of phones had risen to 100 and the phone density to four phones per 100 persons. New phone owners included middle-sized farmers, salesmen (insurance and savings schemes), self-taught village doctors, college students, and graduates who provide tuition, government employees, factory workers, and families who have a family member living and working outside the village. The diffusion of mobile phones in the village followed caste lines, with mainly upper castes buying phones until 2006. By 2010, calling charges had halved in relation to the charges in 2008 and scheduled castes had caught up with the higher castes in phone ownership. In the spring of 2010, only four households out of 158 household surveyed in Janta did not possess a phone. Whereas in 2005 most villagers called from public phones and I could easily film and observe these public calls, by 2010 public calling services were no longer offered in the village as there was no longer a market for them.

When I searched for phone owners in Janta, I was introduced to young men – they are the most mobile household members and are usually also perceived as the principle phones owners. Most women who had acquired personal phones had received them from their fathers, brothers, or husbands. In comparison to Benares, where, according to Doron (2012), young wives' in-laws often confiscate the phones their natal families give them, I never heard of such incidences in rural West Bengal. Such a confiscation would be an extremely rude gesture toward affinal relatives when good relationships between affinal relations are the bedrock of social hierarchies.

Women use phones less than men but, nevertheless, they have access to mobile phones because they are not perceived as private. Seventy-five percent of the survey respondents mentioned that they share their phone. Of the 100 phone calls made from the public phone shops in Janta in 2005, 29% were made by women. By 2011, mobile phones had replaced public phones and women's share of the phone use had grown to 36% based on the phone diaries.

Gendered calling patterns

I asked the survey respondents to list whom they had called that day and what they had talked about. Survey questions have their limitations in that as based on people's recollections and summaries of their calls they provide crude information about phone calls. The survey interview question regarding the call purpose was an open one, and I classified the calls according to the categories the callers gave. Most calls are kept brief because callers want to mimimize the cost; consequently, multipurpose calls were rare and most people could provide one main purpose for the call. Nevertheless, call etiquette usually requires asking how the other person is and what his/her news is, regardless of the main purpose of the call. Many calls had no other purpose than inquiring about the other person's general news. These calls I classified under 'general news'. General news refers to the practice of asking about the other person's news – visiting is also called getting someone's news (khobor neua), which is understood as an integral part of maintaining relationships. My data are representative of the phone use in the village of Janta and the survey results support the patterns I found in Janta.[4]

The survey answers portray women's calling patterns as distinct from men's. The majority of phone owners mentioned calling relatives as the main reason for obtaining their phone. However, in practice men call their friends more than their relatives. Based on the phone diaries, 40% of the calls were men calling their friends, whereas only 1% of the calls entailed women calling their female friends. In the survey (see the Tables 1 and 2), the difference between men and women who talk to their friends was smaller: 25% of men's calls and 12% of women's calls were to friends.

The difference derives from that the phone diaries, unlike the survey, covered no college-attending women – among the women, it is the college students who maintain friendships outside the village. The rest of the women mainly phone their relatives. Men's calls are more often about work or travel than women's calls, which are mainly about discussing the general news or calling for no particular purpose other than to inquire how the other party is. Although these calls serve no specific instrumental purpose, they deepen and strengthen relationships, which may help obtain both emotional and economic support when this is needed. Since women's calls are more limited to their close kin than men's calls, this exemplifies that, in digitally constructed spaces, women are construed as more homebound than men. Jouhki's (2013) research illustrates that mobile phone use follows a similar gendered pattern in the rural areas of the Villupuram district in the state of Tamil Nadu, South India. Here, women use the phone primarily to

Table 1.   Who men called.

Who men called	% of calls
Friends	25
Business partners	21
Unspecified relatives	16
In-laws	10
Work associates	9
Sister	7
Daughter	4
Other villagers' calls	3
Brother	2
Son	1
Wife	1
Teacher	1
	100
	N = 174

Source: Author's survey.

Table 2. Who women called.

Who women called	% of calls
In-laws	19
Husband	17
Mother	12
Friends	12
Unspecified relatives	10
Daughter	10
Son	10
Teacher	5
Father	5
	100
	$N=42$

Source: Author's survey.

contact relatives in their native villages, whereas men prioritize work and business calls and the contacting of friends.

Phones have had a largely unintended effect of diversifying relationships by making it possible for male villagers to maintain relationships with friends outside the village whom they had met at school and at college, whereas women mainly talk by phone with their male relatives. As Castells et al. (2007) have argued, mobile phones influence the nature of the networks. However, far from encouraging individual-centered networks in the vein proposed by Castells et al., mobile phone-enabled networks crucially draw from existing gender relationships and gendered constructions of space. Indian rural men's phone use appears more individualistic than women's because men are freer to establish bonds outside the domain of the home than women, whose gender roles specify that their primary activity domain should be the home. In Granovetter's (1973) terms, men have been able to increase their weak ties (e.g. with acquaintances and friends of friends), whereas phones have helped women mainly to maintain strong kinship bonds. Granovetter (1973) famously argued that jobseekers in Boston found their 'weak' connections to be more useful in the job market than the 'strong' bonds of close friendship and kinship. Granovetter (1983) further argued that poor people's embeddedness in horizontal networks of strong ties with relative equals may serve to isolate them from the kinds of networks that would allow them to have political influence. To understand what it means for women to maintain their strong ties with the help of phones, it is necessary to interpret the changing frames of gendered interaction. I will next describe how women talked about these changes relating them to my observations and then move on to examine how phones mediate these changes.

Changing gender relationships

Women's calling patterns are part of the broader observable changes in the village's gender relationships. These changes include the increase in education for females, women becoming visible in formal politics, and a few high-caste women taking up white-collar jobs. High-caste women have caught up with men in literacy and the younger age groups among the low castes are catching up with the high castes, but the gap between the literacy levels of low and high castes still remains wide. Women represent the village in the *panchayat* and women of all castes regularly attend the *mahila samiti* (women's committee) meetings. The first upper caste village girls (Tilis and Brahmins) graduated from college in the 1990s. Two of them found white-collar jobs (one as a teacher and the other in public administration) in the 1990s; a few women have followed in their occupational footsteps in the following decade. The general attitude toward women opting for white-collar jobs is favorable:

> I have seen that there is no inconvenience. I like service work. It does not harm one's health. A service job provides one with an income of one's own and allows one to buy what one wants. Without a job we have to ask our husbands for money. (A Tili woman)

The acceptance of female college students and women in white-collar jobs comes easily, even by men, as the majority of the villagers have little or no first-hand experience of such drastic changes. The great majority of village women is not employed outside the home. In general, high-caste women do not work outside their house or move about alone outside their neighborhood. Upper caste married women rarely move outside their own neighborhood and village alone, although a Tili woman may travel alone to see a doctor or visit her parental home if there is an emergency. They usually leave the village – to go shopping, to see a movie in Vishnupur, and to visit their parents – with their husbands.

A married woman moving on her own outside her home is under scrutiny as, potentially, it is a threat to the family honor. The higher the caste and class, the more restricted a woman's opportunities to move outside her house. Yet, most women do not live under constant surveillance and prohibitions. Family members usually share an understanding of the nuances of women's going out. Most women have the freedom to move about in their own caste neighborhood and are allowed to go outside the village accompanied by neighbors and in the presence of family men.

When asked about the concrete manifestations of changes, most women say that instead of radical changes, they have experienced subtle reforms in family life to which the increase in the general standard of living has contributed. A higher standard of living, they say, enables one to be a good family person, to live happily as part of a small family where women can express their opinions, concerns, and wishes. The goals of freedom translate into husbands being more attentive to their wives' wishes – by taking them to the cinema, shopping, and on picnics. As a result of the increasing prosperity, husbands are now better able to fulfill their wives' fascinations (*shoks*): women can dress better, eat better, and decorate their houses. Women, who at most have only realized subtle changes in their lives, express a determination to provide more major changes for their daughters: better education, less hard work, new job options, and, later, marriage to good families along with substantial dowries. Instead of rejecting the traditional gender and kinship code of conduct, women seek to reform it.

Political activism has been important in making women conscious of their rights in culturally meaningful ways although opening up new political spaces for women has not led to drastic changes in economic power structures. The scope of the *panchayats'* power is limited and state policies toward women have been highly ambiguous in West Bengal. While involving women in its activities and state administration, the CPI(M) party carried out land reform without considering women as land owners until most of the land had been distributed. Land mainly gets passed through the male line through inheritance and ideas of the home as the women's place prevent women from participating in market activities such as running small-scale businesses. Only few middle-class women have benefited from women's rights discourses economically, as these discourses have encouraged them to get a college education which has made it possible for them to find public sector jobs. Home and the domain of the household has become the main arena for pursuing women's rights in rural West Bengal, while the idea of the public, economic sphere as the men's domain remains largely unchallenged. Female teacher from Janta assessed the role of phones for women in relation to many barriers women experience: 'Women here have many problems and phones do not solve all of them.'

Mediation of gendered space through mobile phones

An often noted feature of mobile phones is that they help blur boundaries of social spaces (Katz, and Aakhus 2002; Ling 2004; Horst and Miller 2006; Donner 2009; Sey 2011). In rural India, the introduction of phones offered women a new, unobtrusive avenue to extend their contacts and space without moving out of their neighborhood. Similarly to television, translocal communication with the help of phones has helped bring the outer world into the women's sphere. This has changed their contact with close relatives although there has not been a clear linear change of calls replacing visits. The occasions when women can visit their natal families have never been numerous: women are in charge of the household work and if there is no one to look after the children and take care of the household, it is difficult for them to leave on visits. Women do, however, visit their parents' home during one or more annual religious festivals. Women who live close to their natal families are able to visit more often. One woman, for instance, called her parents daily and also visited them weekly. The ability to call may at times replace intended visits, but it can also strengthen the relationship and encourage more visiting than would have been possible without phones. Of course, both the ability to call and visit as one pleases crucially depend on whether one can afford these activities. Most people commented that a short phone call saves time and money compared to having to travel by bus to deliver important news. Thanks to the phones, women are better connected with their natal families, which, for most women, are a major source of support. Even low-income families can afford to use phones to convey news about emergencies. Just a decade ago, women could be facing food scarcity or were mistreated in their husband's house for years before the news reached their parents. Today, the exchange of news is so intense that, for instance, news of loss caused by a hailstorm reached women's parents within a few days, although they did not call their parents directly.

Phones are also used to discuss marriage arrangements.[5] People inform one another about arranged marriages, ask information about potential brides and grooms, ask advice about marriage offers, and deliver news of the acceptance or rejection of marriage proposals. The most decisive part of the marriage arrangement is when the groom's party comes to see the potential bride. The girl and boy can see each other and afterwards give their opinion on the marriage proposal, although marriages are sometimes arranged without taking the bride's and groom's views into consideration. Phones help arrange these meetings and make it possible for the potential grooms to see more brides than they could have before the arrival of the phones. Phones have also made it easier to express opinions on potential brides and grooms – rejections no longer need to be conveyed in person.

It is traditionally the father's responsibility to look for suitable brides and grooms. Mothers cannot go and find potential brides for their sons and most certainly not grooms for their daughters, but they nevertheless participate in marriage arrangements by finding out about suitable boys and grooms though their networks. The introduction of the working phone system to the region has increased women's role in marriage negotiations. For instance, after a father of a prospective bride had invited several mismatches to the house to see his daughter, the neighboring women made a few phone calls and quickly arranged a meeting with a promising candidate: this potential groom owned some land, had a side business, and was known to be good natured and hardworking.

Phones also help women to stay in touch with their families when they physically move outside the home. The few village women who go to college or have a service job outside the village always carry their personal phones. They use these phones to inform their home about their schedules and possible delays in commuting from work and to monitor how things are at home. I did not hear women express that they felt calls from home as unwanted surveillance. Women value the possibility to call for help if they face problems, such as missing buses, accidents, break downs of vehicles, jams, and demonstrations on the road. As a female college student relates:

> I travel to college by bus, so people can call me to tell me if the bus is not going to come. And if the last bus does not come, I can call home. The ability to call gives me mental courage. If I face any inconvenience outside the home, they [the family] will come and get me.

Phones have helped to change the meaning of the outside sphere for her, conflating it with the home sphere while, simultaneously, giving her courage and thereby influencing her subjectivity.

Mediation of speech contexts by phones

Phones were a novelty in rural West Bengal in the sense that they offered people new possibilities to choose the context of their talk. Opportunities for private conversations in the village are limited but people do try to maneuver in order to not to share all their discussions with the neighbors and the extended family. Mobile phones offer the possibility to move away so that fewer people are within hearing distance. Tacchi, Kitner, and Crawford (2012) similarly observed that women valued their newly found ability to talk by phone without everyone in the household or vicinity hearing their converstations in the Telangana Region of Andhra Pradesh. In Janta, a few young daughters-in-law confessed that they usually called their natal homes when the in-laws were not at home.

Before the establishment of a working phone system in rural West Bengal, young women were not supposed to meet or contact their natal families for one year after their marriage – this practice was rationalized by the need to help newly wed wives adjust to their in-laws' households. Older women proudly told me about this challenging transition – moving to live with strangers. The distance between houses was experienced as contributing to the honor and status of the houses, while good relationships with in-laws are highly valued. Thanks to phones, it has now become commonplace for newly wed women to call their parents – even daily. A newly wed wife in the village could be completely occupied by her mobile phone even right after her marriage. 'There she sits, holding her phone, waiting for a call, or talking over the phone with her natal family', the older women described a new wife in the neighborhood, fearing that the phone might indeed hamper her intergration into her in-laws' family lives. In some families in-laws encourage their daughter-in-law to call her natal family regularly, whereas in other families a young wife has to cope with the in-laws' often tacit reluctance to allow her to call. A young wife replied to my question whether her in-laws minded her using the phone:

> They do not mind because they need not know about my phone use. My husband gives me the money for calls. I usually call when my father-in-law or mother-in-law are not at home. If they are at home and I need to call, I go to the attic to make the call.

In this family, a young wife's calling practice has strengthened the couple's unity in relation to the broader family unit, as the husband is funding the wife's calls without his parents knowing. Since men are usually in charge of the household economy and only a few women have their own earnings, women's ability to use phones requires support from a male family member. And many husbands, brothers, and fathers do fund women's calling when they can afford it. In most families, phones are perceived as a collective family resource and men are motivated to contribute to the well-being of the women in their families by allowing women to use the phones. It would, therefore, not do justice to measure women's phone use solely through the notion of the Euro-American personhood as an ability to use personal phone privately, although women increasingly have personal phones – which they mostly receive as presents from their male kin.

The importance attached to relationships with in-laws has clearly helped women to gain an access to phones. By calling their parents frequently, women have not adopted a completely new practice, but have instead strengthened the relationships between kin groups which were already valued as important. However, the greater communication density that phones enable is new. At the same time, a woman's ability to call also reflects the position she has been able to carve in her in-laws' house, as well as the economic standing of the household. When women are able to call freely it signifies that they enjoy a good relationship with their husbands and/or in-laws and that the household is wealthy enough to allow calling.

A woman from Janta who, over the phone, advised her daughter to disobey her mother-in-law is an example of how communication with natal relatives can have subversive elements. Her daughter had married into a well-to-do household where she was responsible for all the housework. The daughter was happily married in that she was well off, but her workload exhausted her. Usually women share tasks more equally, although mother-in-laws tend to be in position of power. The mother advised her daughter to simply refuse to do the excess work in her in-laws' house. She feared that if the daughter kept obeying, her workload would grow unbearable. Following her mother's advice, the daughter successfully refused extra chores. Without a phone, the chances for this conversation would have been limited, because the mother would have usually only met her daughter when surrounded by her in-laws. Another phone conversation between a mother and daughter entailed a highly detailed conversation about farming options, which women rarely discussed in public because farming decisions are regarded as part of the male domain. Phones offer women a channel to express unconventional ideas and exert their will through networking by offering them a chance to speak to only one listener at a time if they so choose.

Phones also make it possible to maintain other defiant and secret contacts. A young woman, for instance, – unconventionally – discussed her own dowry arrangements with her married sister, who lived outside the village. Young college students can maintain relationships with boyfriends and girlfriends through personal phones, although non-marital relationships are still rare in the region. The survey included a young woman whose boyfriend had purchased a phone for her and a married woman who used her personal phone to communicate with her lover. Nevertheless, phones are not considered a major threat for the marriage system, which explains why the discourse on the disruptive effect of phones was relatively weak in the region during my research periods. I tried to ask people about the harmful effect of phones, but they did not complain about phones spoiling social relationships. I witnessed romantic liasons, which led to love marriages, in Janta even before people started to use phones, but there has been no linear shift, no direct change from arranged marriages to love marriages. Arranged marriages are in fact having an increasing economic impact in the region because of the growing demands for dowries (Tenhunen 2008b). Love marriges, which occur occasionally, are not experienced as highly disruptive because they usually support caste endogamy and resemble arranged marriages in that the parents or guardians of the couples usually get to negotiate the dowry and set the marriage date.

Conclusions

Whereas in many parts of India women's mobile phone use has been experienced as a threat to the marriage system (Grodzins Gold 2009; Doron 2012; Jouhki 2013; Kärki 2013), in rural West Bengal the marriage system and the ensuing hierarchical relationships between kin groups have encouraged and legitimated women's mobile phone use. Women do not cut their ties to their natal families at the time of marriage and the maintenance of relationhips between kin groups is a highly valued activity, which explains how women have carved a space to increase their contacts with their natal families. This increased communication can, in turn, provide women with both economic and mental supports in a wide variety of situation, even against the will of their in-laws. It is a tangible proof of the changes caused by the women's ability to use mobile phones that it has become a common place for newly wed wives to stay in touch with their parents over the phone right after their marriage, whereas just a decade ago I observed that contact between kin groups was avoided for a year after the marriage. Moreover, it has become acceptable for women to own personal phones although when phones first became available men were usually regarded as the main owners.

I have illustrated how the physical qualities of phones help strengthen the multiplicity of discourses by mediating relationships. First, they enable translocal communication, helping callers to transgress social boundaries: in other words, women benefit due to the reconstruction of the meaning of the home and the outside world, which phones have enabled. Second, phones give callers new possibilities to choose the context for their speech and engage in critical and unconventional discourses. Instead of drastic improvements or changes, for instance in economic power relationships, the positive impacts of women's phone use appear subtle and ambiguous: most calls are about the slight redefinition of the home boundaries. Mobile telephony amplifies dominant discourses in that calling patterns are gendered, but by helping to blur cultural boundaries and providing unconventional speech contexts, it also creates spaces for agency and critical discourse. As Wajcman (2002) argues, ICTs do not offer simple technological fixes to social problems but they are part of social changes through how technologies are produced and used socially. Detailed attention to the multiple uses and influences, in turn, can help to create development interventions and policies which take into account the multiplicity of actors and ongoing social processes.

Notes

1. The survey covered respondents from the following villages: Dhengasole, Satmouli, Ghugimura, Lego, Chatramore, Parairy, Chandabila, and Pamua.
2. This article also draws from my earlier work in Janta. I lived in the village for 10 months in 1999–2000 and returned to Bankura for 4 more months of fieldwork in 2003 and 2004. My earlier research in Janta focused on gender, politics, and exchange relationships (Tenhunen 2003, 2008a, 2008b, 2011). The bulk of the earlier materials consists of 76 taped interviews with villagers and a complete household census of the village, but I also gathered many of my insights into village life through observation, participant observation, and chatting.

3. Tenhunen (2011) focuses on the use of mobile communication in politics in rural West Bengal.
4. However, the survey is not statistically representative of the Bankura region.
5. Survey did not include calls for marriage arrangements because it was not carried out during the marriage season. I, however, had the chance to observe calls for marriage arrangements in the village during the marriage season.

References

Altheid, David, and Robert Snow. 1988. "Toward a Theory of Mediation." In *Communication Yearbook 11*, edited by James A. Andersen, 194–223. New York: Routledge.

Archambault, Julie Soleil. 2010. "La fièvre des téléphones portables: Un chapitre de la 'success story' mozambicaine?" *Politique Africaine* 117 (1): 83–105.

Barendregt, Bart. 2008. "Sex, Cannibals, and the Language of Cool: Indonesian Tales of the Phone and Modernity." *The Information Society* 24 (3): 160–170.

Boellstorff, Tom. 2008. *An Anthropologist Explores the Virtually Human*. Princeton, NJ: Princeton University Press.

Bourdieu, Pierre. [1980] 1992. *The Logic of Practice*. Cambridge: Polity Press.

Burke, Jason, and Manoj Kumar. 2012. "Indian Village Bars Women from Using Mobile Phones." *The Guardian,* December 5. Accessed July 17. http://www.guardian.co.uk/world/2012/dec/05/india-village-bars-women-mobile-phones

Castells, Manuel, Mireia Fernández-Ardèvol, Jack Linchuan Qiu, and Araba Sey. 2007. *Mobile Communication and Society: A Global Perspective.* Oxford: MIT Press.

Chib, Arul, and Vivian Hsueh-Hua Chen. 2011. "Midwives with Mobiles: A Dialectical Perspective on Gender Arising from Technology Introduction in Rural Indonesia." *New Media & Society* 13 (3): 486–501.

Cockburn, Cythia. 1983. *Brothers: Male Dominance and Technological Change*. London: Pluto.

Collier, Jane, and Sylvia Yanagisako, eds. 1987. *Gender and Kinship: Essays toward a Unified Analysis*. Stanford, CA: Stanford University Press.

Donner, Jonathan. 2009. "Blurring Livelihoods and Lives: The Social Uses of Mobile Phones and Socioeconomic Development." *Innovations: Technology, Governance, Globalization* 4 (1): 91–101.

Doron, Assa. 2012. "Mobile Persons: Cell Phones, Gender and the Self in North India." *The Asia Pacific Journal of Anthropology* 13 (5): 414–433.

Encheva, Kameliya, Olivier Driessens, and Hans Verstraeten. 2013. "The Mediatization of Deviant Subcultures: An Analysis of the Media-Related Practices of Graffiti Writers and Skaters." *MedieKultur* 29 (54): 8–25.

Fischer, Claude S. 1992. *America Calling: A Social History of the Telephone to 1940*. Berkeley: University of California Press.

Fruzzetti, Lina. [1982] 1990. *The Gift of a Virgin: Women, Marriage, and Ritual in a Bengali Society*. New Delhi: Oxford University Press.

Granovetter, Mark. 1973. "The Strength of Weak Ties." *American Journal of Sociology* 78 (6): 1360–1380.

Granovetter, Mark. 1983. "The Strength of Weak Ties: A Network Theory Revisited." *Sociological Theory* 1: 201–233.

Grodzins Gold, Susan. 2009. "Tasteless Profits and Vexed Moralities: Assessments of the Present in Rural Rajasthan." *Journal of the Royal Anthropological Institute* 15 (2): 365–385.

Hjarvard, Stig. 2008a. "The Mediatization of Society: A Theory of the Media as Agents of Social and Cultural Change." *Nordicom Review* 29 (2): 105–134.

Hjarvard, Stig. 2008b. "The Mediatization of Religion: A Theory of the Media as Agents of Religious Change." *Northern Lights 2008. Yearbook of Film & Media Studies* 6 (1): 9–26.

Horst, Heather, and Daniel Miller. 2006. *The Cell Phone: An Anthropology of Communication*. Oxford: Berg Publishers.

Horst, Heather, and Daniel Miller. 2012. *Digital Anthropology*. Oxford: Berg Publishers.

International Telecommunication Union. 2010. "The World in Figures. ICT Fact and Figures." Accessed July 23. http://www.itu.int/ITU-D/ict/material/FactsFigures2010.pdf

Jouhki, Jukka. 2013. "A Phone of One's Own? Social Value, Cultural Meaning and Gendered Use of the Mobile Phone in South India." *Journal of the Finnish Anthropological Society* 38 (1): 37–58.

Kärki, Jelena. 2013. "'If My Daughter Runs Away, I Will Drink Poison' an Anthropological Study of Child Marriage in North Indian Villages." Unpublished master's thesis manuscript.

Katz, James, and Mark Aakhus, eds. 2002. *Perpetual Contact: Mobile Communication, Private talk, Public Performance*. Cambridge: Cambridge University Press.

Latour, Bruno. 1999. *Pandora's Hope. Essays on the Reality of Science Studies*. Cambridge, MA: Harvard University Press.

Ling, Richard. 1998. *"She Calls, [but] It's for Both of Us You Know": The Use of Traditional Fixed and Mobile Telephony for Social Networking Among Norwegian Parents*. Kjeller: Telenor Forskning og Utvikling.

Ling, Richard. 2004. *The Mobile Connection: The Cell Phone's Impact on Society*. San Francisco, CA: Morgan Kaufman.

Lohan, Maria, and Wendy Faulkner. 2004. "Masculinities and Technologies: Some Introductory Remarks." *Men and Masculinities* 6 (4): 319–329.

Mazzarella, William. 2004. "Culture, Globalization, Mediation." *Annual Review of Anthropology* 33: 345–367.

Moyal, Ann. 1992. "The Gendered Use-of the Telephone: An Australian Case Study." *Media, Culture and Society* 14 (1): 51–72.

Oakley, Ann. 1974. *The Sociology of Housework*. London: Martin Robertson.

Ortner, Sherry B. 1989. *High Religion: A Cultural and Political History of Sherpa Buddhism*. Princeton, NJ: Princeton University Press.

Planning Commission. 2012. "Press Note on Poverty Estimates, 2009–10." Accessed July 26. http://planningcommission.nic.in/news/press_pov1903.pdf

Reckwitz, Andreas. 2002. "Toward a Theory of Social Practices: A Development in Culturalist Theorizing." *European Journal of Social Theory* 5 (2): 243–263.

Sahlins, Marshall. [1985] 1987. *Islands of History*. Chicago, IL: University of Chicago Press.

Schatzki, Theodore R., Karin Knorr-Cetina, and Eike von Savigny, eds. 2001. *The Practice Turn in Contemporary Theory*. London: Routledge.

Schneider, David. 1980. *American Kinship: A Cultural Account*. Chicago, IL: University of Chicago Press.

Schulz, Winfried. 2004. "Reconstructing Mediatization as an Analytical Concept." *European Journal of Communication* 19 (1): 87–101.

Sey, Araba. 2011. "'We Use It Different', Making Sense of Mobile Phone Use in Ghana." *New Media and Society* 13 (3): 375–390.

Shove, Elizabeth, Mika Pantzar, and Matt Watson. 2012. *The Dynamics of Social Practice: Everyday Life and How it Changes*. Los Angles, CA: Sage.

Strathern, Marilyn. 1992. *After Nature: English Kinship in the Late Twentieth Century*. Cambridge: Cambridge University Press.

Tacchi, Jo, Kathi Kitner, and Kate Crawford. 2012. "Meaningful Mobility: Gender, Development and Mobile Phones." *Feminist Media Studies* 12 (4): 528–537.

Telecom Regulatory Authority of India. 2012. *Annual Report 2011–12*. New Delhi: Telecom Regulatory Authority of India.

Tenhunen, Sirpa. 2003. "Culture and Political Agency: Gender, Kinship and Village Politics in West Bengal." *Contributions to Indian Sociology* 37 (3): 495–518.

Tenhunen, Sirpa. 2008a. "Mobile Technology in the Village: ICTs, Culture, and Social Logistics in India." *Journal of the Royal Anthropological Institute* 14 (3): 515–534.

Tenhunen, Sirpa. 2008b. "Gift of Money: Rearticulating Tradition and Market Economy." *Modern Asian Studies* 42 (5): 1035–1055.

Tenhunen, Sirpa. 2011. "Culture, Conflict and Translocal Communication: Mobile Technology and Politics in Rural West Bengal, India." *Ethnos* 76 (3): 398–420.

Wajcman, Judy. 1991. *Feminism Confronts Technology*. Cambridge: Polity.

Wajcman, Judy. 2002. "Adressing Technological Change: The Challenge to Social Theory." *Current Sociology* 50 (3): 347–363.

Wallis, Cara. 2011. "Mobile Phones without Guarantees: The Promises of Technology and the Contingencies of Culture." *New Media & Society* 13 (3): 471–485.

Supply-and-demand demographics: dowry, daughter aversion and marriage markets in contemporary north India

Patricia Jeffery

School of Social and Political Science, Sociology & Centre for South Asian Studies, University of Edinburgh, Chrystal Macmillan Building, Edinburgh, UK

The gendered character of India's fertility decline has attracted considerable academic attention. In this paper, I offer a critique of the arguments of some demographers about the linkages between dowry, daughter aversion and the marriage squeeze that predict that increasing shortages of marriageable women will result in declines in dowry. I argue that such economistic readings seriously oversimplify the complexities of marriage arrangement 'on the ground' in contemporary India. Further, whilst one aspect of dowry might relate to the supply and demand of brides and grooms, dowry and daughter aversion are not simply outcomes of demographics alone. First, marriage migration is crucial in understanding daughter aversion. Second, dowry is not just a matter of marriage and kinship practices. Dowry is a polyvalent institution that also connects with conspicuous display in status competition in a hierarchical society and with people's rising aspirations to possess consumer goods within the wider context of contemporary India's rapidly changing political economy. Crucially, marriage migration, status competition and consumerism do not necessarily push in the same direction as the demographics of the marriage squeeze might imply when it comes to dowry and daughter aversion.

Introduction

One feature of India's demographic profile that has attracted demographers' attention for many decades has been its masculine sex ratio. This interest received a marked impetus when India's 2001 census indicated that child sex ratios (CSR, the sex ratio of children aged 0–6) had become significantly more masculine since the 1991 census. CSRs had been becoming more masculine throughout India since the 1970s and at an accelerating pace since the mid-1980s. Fertility has been declining throughout India, with its onset later and slower in the north and north-west than in the south. Similarly, processes associated with sex bias are not spatially random (Guilmoto and Attané 2007).

My research in rural Uttar Pradesh since the early 1980s has focused on gender issues, including those related to these demographic changes and their linkages to family life and the wider economy. This paper, however, is primarily based on my exploration of various relevant research literatures, including those of demographers. Whilst my argument draws only indirectly on my own research experience, it is nevertheless heavily informed by it – most particularly by my feeling that the assertions made by some demographers about the connections between dowry, 'daughter aversion' and India's increasingly masculine CSRs fail to make sense of processes on the ground.

Several demographers suggest that India's demographic profile will result in a 'marriage squeeze', in which the relative sizes of cohorts of potential brides and potential grooms imply that brides will be in short supply. Guilmoto, for instance, cautiously notes that 'while discrimination against unborn girls today is a dismal reflection of the status of women, sex imbalances may also lead tomorrow to the potential disruption of marriage systems set off by the unavoidable shortage in prospective brides' (2012, 78). He does not, however, align himself with P. N. Mari Bhat and Monica Das Gupta and their colleagues, who have outlined some rather optimistic but questionable connections between dowry, daughter aversion and the marriage squeeze (Bhat and Halli 1999; Das Gupta and Li 1999; Das Gupta et al. 2003). When grooms are scarce, their argument runs, the marriage squeeze works against women: dowries increase because larger numbers of brides seek scarce grooms, and parents become more daughter averse. In light of the recent decline in fertility and the use of sex-selective abortions, Bhat and Das Gupta predict that potential brides will become scarce, the marriage squeeze will begin to operate against potential grooms – and parents will become less daughter averse because dowry will decline in importance.[1] This deployment of an economistic framework in the realm of kinship and gender politics rang danger bells for me, however – and my disquiet impelled me to think through and elaborate the reasons for my disquiet. This paper, then, is the result of my reflections on what might be termed 'supply-and-demand demographics'.

I shall argue that, however, plausible Bhat and Gupta's predictions might seem at first glance, they cannot withstand close scrutiny. Even within its own terms, their economistic marriage squeeze model can be faulted because it seriously oversimplifies the complexities of marriage arrangements in India. Examining how marriages are arranged 'on the ground' provides a more fruitful means of interrogating the links between dowry, daughter aversion and the marriage squeeze.

Further, whilst one aspect of dowry might relate to the supply and demand of brides and grooms, dowry and daughter aversion are not simply outcomes of demographics alone. As I shall also argue, marriage migration is crucial in understanding daughter aversion. But nor is dowry just a matter of marriage and kinship practices either. Put somewhat differently, gender politics are everywhere. There is no escaping them. And it is vital to see dowry as a polyvalent institution that also connects with conspicuous display in status competition in a hierarchical society and with people's rising aspirations to possess consumer goods within the wider context of contemporary India's rapidly changing political economy. Crucially, marriage migration, status competition and consumerism do not necessarily push in the same direction as the demographics of the marriage squeeze might imply when it comes to dowry and daughter aversion.

Gender bias in a time of fertility decline

This paper is mainly concerned with areas of north and north-west India (e.g. Punjab and Haryana) that both historically and currently are notorious for having the most masculine CSRs in India (Dyson 2001; Arokiasamy 2004, 2007; Visaria 2004b) – including what Oldenburg (1992) notably termed the 'Bermuda triangle for girls'. But there has also been spatial diffusion of increasingly masculine CSRs encompassing contiguous areas elsewhere in northern India. Indeed, the 2011 census indicates that CSRs have plateaued in Punjab and Haryana, but CSRs have become more masculine in several other northern states (e.g. UP, Bihar, Rajasthan, and Uttarakhand). In northern India, declining fertility is associated with couples' increasing efforts to affect the gender balance of their children (Guilmoto and Attané 2007; Guilmoto 2008), an important downside of what Agnihotri (2000, Chapter 8) terms 'prosperity optimism'. The implications of this demographic shift are already being researched by sociologists and anthropologists (e.g. John et al. 2008; Purewal 2010; Larsen 2011).

Until the 1980s, masculine CSRs were basically explicable in terms of differential care of girls and boys that led to higher rates of infant and child mortality among girls. Even today, in the northern states, female disadvantage sets in very soon after the neonatal period (when boys are biologically more at risk) (Arokiasamy and Gautam 2008). Since the mid-1980s, however, new technologies (first amniocentesis, latterly ultrasound) have become widely available and affordable. These technologies were initially developed as the means of detecting foetal abnormalities in utero but they are also capable of determining foetal sex. As many commentators have noted in respect of technological innovation in general, new technologies do not enter neutral fields: and in the context of widespread gender bias in India, ultrasound is now widely used for sex determination followed by sex-selective abortion (despite legislative interventions outlawing such usage). Discussions of the incidence and social distribution of sex-selective abortions in India are necessarily based on estimates, however, since reliable information is scarce (e.g. Agnihotri 2001a; Arnold, Kishor, and Roy 2002; Patel 2007; Guilmoto 2009). Arnold et al. (2002) estimated that sex-selective abortions accounted for over 100,000 out of a total of 1.3 million abortions per annum by the turn of the millennium. More recently, Kulkarni (2012) has calculated that around 400,000 sex-selective abortions are performed each year, accounting for the major part of the female deficit in the CSRs. Masculine sex ratios at birth (SRBs) are another indication of sex-selective abortions and SRBs in India during the 2000s have hovered around 900 (even dipping to 880 in 2003–2004). Normal SRBs are around 930–960 girls per 1000 boys (equivalent to 104–107 males per 100 females.[2]

Some commentators argue that there is a substitution effect, with neglect and differential care being replaced by sex-selective abortion (e.g. Goodkind 1996). By contrast, Das Gupta and Mari Bhat (1997) talk of an intensification effect, with increased discrimination against girls when fertility declines faster than the desired number of sons; they considered that excess female child mortality accounted for four times more missing girls than sex-selective abortion in the mid- to late 1990s, although they predicted that sex-selection would play an increasingly significant role. Bhat and Francis Zavier (2003) rephrase intensification as a 'son preference' effect that can more effectively be put into practice because of newly available technology. Similarly, Sudha and Irudaya Rajan (1999, 2003) consider that girls face 'double jeopardy' from neglect compounded by sex-selective abortion and Agnihotri (2001b, 2003) considers that sex-selective abortion can coexist with the continuation of discrimination against those girls who are born.

Indeed, educated, wealthy urban residents are apparently more likely than poorer rural residents to practise sex-selective abortions, but they discriminate less against girls they allow to be born; by contrast, excess female child mortality is more marked for children of poorer and uneducated mothers (Agnihotri 2003; Arokiasamy 2004, 2007; Attané and Guilmoto 2007). It also seems that the economically advantaged 'pioneer' sex-selective abortions and the practice gradually seeps down the class hierarchy within localities (Guilmoto and Attané 2007; Guilmoto 2008). Consistent with this is the spread of masculine

CSRs to new areas in northern India as well as in the south, beginning in the urban centres and gradually spreading to rural areas (Agnihotri 2003).

SRBs are most masculine for Sikhs and Jains, followed by Hindus (Bhat and Francis Zavier 2007; Guilmoto 2008). Probably because of improvements in Scheduled Castes' economic position, masculine CSRs among them are intensifying, so narrowing the gap between them and the general population (Bhat 2002a, 2002b; Bhat and Francis Zavier 2003; Siddhanta, Agnihotri, and Nandy 2009). Broadly, Muslims reflect the overarching regional patterns, yet within regions Muslims have less masculine CSRs than their neighbours (Guilmoto 2008). Muslims both express less preference for sons and rarely practise sex-selective abortion (Bhat and Francis Zavier 2003, 2007). Further, their children (including girls) have a mortality advantage that cannot be readily explained by the differentials in wealth and education that seem to relate to CSR differences for caste Hindus and Scheduled Castes (Bhalotra, Valente, and van Soest 2010a, 2010b).

Demographers generally consider that fertility will continue to decline in India, with a gradual convergence between the different regions. There is less agreement, however, about the gender bias in this process. Some sources suggest that the SRB has begun to plateau (Bhat 2002b; Bhat and Francis Zavier 2003) and that son preferences will decline (Visaria 2004a). Others, however, have predicted that India's sex ratios will continue to become more masculine (Das Gupta and Mari Bhat 1997; Mayer 1999) and that son preferences will be manifest for some decades, especially if the social and economic root causes are not removed (Guilmoto and Attané 2007). In particular, the north-west is likely to have masculine CSRs before they might begin to plateau (Kaur 2007; Das Gupta 2009). Much of the rest of north India, where fertility decline has developed later, will probably have worsening CSRs, at least for a time (Guilmoto 2009).

Dowry and daughter aversion

Dowry has a long history in northern India and much of the vast literature exploring the reasons behind the marked gender bias in the fertility transition focuses on parental fears about the financial implications of having to provide daughters with dowries. Most writers focus on the items that are transferred shortly before a marriage (e.g. at engagement ceremonies) and especially at the time of a marriage. These include clothing and jewellery given by the bride's family to the bride herself and other items (sometimes termed 'groom-price') that are destined for the groom and his relatives (e.g. household goods, vehicles, cash, clothing, jewellery). It is important to emphasise, however, that local understandings regard dowry as inseparable from other outlays faced by the bride's family: particularly hospitality during the wedding festivities and the patterns of gift-giving from a bride's natal kin to her affines that are initiated at the wedding and endure throughout the marriage and beyond. The dowry is usually the most substantial of these transfers, but the continuing giving sustains a bride's links with her natal kin. When a married daughter visits her parents she should return to her husband's home with gifts, and festivals, harvests and family events (births of sons, marriages, etc.) are also marked by this unilateral flow of goods, which may comprise clothing, jewellery, livestock, and/or portions of grain crops. Parents who are unable to fulfil these obligations may stop inviting their daughter for visits or attending weddings in her in-laws' household. In northern India, these patterns of giving were historically particularly associated with upper-caste Hindus and framed in terms of *kanya-dān* (gift of a virgin, in which the bride's parents give the bride with no expectation, or indeed option, of any return), and its associated implications of hypergamy and status asymmetry between wife-givers and wife-takers (Sharma 1984; Srinivas 1984; Kumari 1989; Jeffery and Jeffery 1996; Basu 2001, 2005a).

I am not aware of any systematic national- or even regional-level studies with time-series data on marriage payments. Furthermore, small-scale studies have used different definitions of dowry. This means that we cannot draw firm conclusions on how dowry has changed. But we do have a fairly good idea of how people themselves think that dowry has changed. Since the publication of the report on The Status of Women in India (ICSSR 1975), a standard narrative about dowry has attained a common-sense status in academic and popular media accounts alike. The nearest thing to an overall account is provided by All India Democratic Women's Association based on some 10,000 interviews around the country (AIDWA 2003). Its findings lend support to the common-sense view. People surveyed dated several major changes from about the mid-1980s: dowry appeared in regions where it was virtually unknown and spread within regions to groups that previously did not give dowry. Dowry has apparently increased in quantity (although whether in real terms and in relation to incomes cannot be determined). Certainly, new items should be included and cash is now seen as crucial. Increasingly, dowry is associated with more overt and aggressive demands for specific items made by the groom's family, and brides whose parents fail to comply are said to be at risk of increasing levels of harassment (and even murder). According to AIDWA (2003), women often saw dowry as vital for their own security. There are also indications that more educated men command larger dowries, dowries are largest among the most wealthy, poorer families often incur serious debts to provide dowries, and that the continued giving over the years is viewed as increasingly burdensome.

These findings concur with numerous other sources. Srinivas (1984, 19) considers that dowry had become obligatory and already by the 1980s was characterised by 'asymmetry [i.e. hypergamy], uncertainty and unpredictability' because of demands even after the marriage and because weddings themselves had become occasions for conspicuous spending and public claims for status. Abdul Aziz attributed the element of compulsion in dowry to a shift from the 'normal eligible bachelor to a "fancy" product' (with English education, formal sector job, etc.); further, adolescent daughters are 'perishable commodities' who must be married as soon as possible because of fears that they will be 'dishonoured' (Aziz 1983, 604). Similarly, Sambrani and Sambrani (1983, 602) noted the 'expropriatory characteristics' of dowry and the 'virtual auction of the eligible men to the highest bidders' with dowry escalation fuelled by the pressure on girls' parents to accept a good match as soon as one materialises. Twenty years later, Kaur (2007) reports that people in Haryana connect seeing a daughter as a burden with the need to provide a dowry, a view that is paralleled in my own fieldwork (see e.g. Jeffery and Jeffery 1996, 69–97). Writing about Muslims in central India, Vatuk (2007) reports that they view dowry as a cancer or running sore (*jahez kā nāsūr*) that has been adopted quite recently in imitation of Hindu practices: people associated it with a decline in marriages between close relatives.

It is, then, likely that dowries have increased in size and have spread to new areas and groups, and plausible that the pace of change has increased since the 1980s and especially since the economic reforms of the early 1990s. In any case, people's perceptions of dowry escalation, increasing demands, and harassment are probably more salient for understanding how daughter aversion might relate to dowry. Here I shall focus on arguments about the relationships between changes in dowry and the 'marriage squeeze'. How (if at all) are dowry practices linked to demographic change and daughter aversion? Does dowry-giving reflect a surplus of marriageable young women – and will it decline if the marriage squeeze reverses?

Dowry and the marriage squeeze

In India, men usually marry women a few years younger (say 2–10 years: age differences are generally lowest in the north). Thus assessments of the 'marriage squeeze' necessitate comparing the sizes of relevant age cohorts. During the early twentieth century, high levels of child mortality meant that younger age cohorts were generally similar in size to those somewhat older, although sex differentials in mortality created some 'surplus' of potential grooms. Since 1947, however, child mortality declined earlier than fertility: younger cohorts became larger, although the 'surplus' of females was not as large as it might have been, given higher levels of female child mortality. Additionally, improvements in adult women's survival reduced levels of widowerhood and thus of men available for second marriages (Bhat and Halli 1999).

According to the arguments of Bhat and of Das Gupta and their colleagues (Bhat and Halli 1999; Das Gupta and Li 1999; Das Gupta et al. 2003) linking dowry to the 'marriage squeeze', the resulting 'shortage' of marriageable men has meant that girls' parents feel compelled to respond to escalating dowry demands in order to ensure that their daughters can marry. Moreover, they predict that the demographic parameters that they consider responsible for dowry escalation are set to change – with knock-on effects on dowry and daughter aversion. When fertility declines (and especially when the decline is rapid), younger cohorts become smaller than older ones. In some regions, the resulting 'shortages' of young women are exacerbated by excess female child mortality and sex-selective abortion. And, so the argument runs, dowry will begin to wither. Bhat and Halli (1999) argue that a shortage of potential brides increased after about 1991. Rising ages of marriage for men might compensate (but only partially) for this reversal in the 'marriage squeeze', so gradually the 'price' of grooms on the marriage market would probably reduce. Similarly, Das Gupta and others suggest that 'the surplus of men that we can expect for birth cohorts after 1980 means that there is hope that dowry inflation will taper off' (Das Gupta and Li 1999, 363) and that

> [m]any studies indicate that the surge in dowry payments in India is related to an unusual con-
> figuration of demographic forces, which are no longer at play. Population projections indicate
> that a shortage of brides and brideprice will begin to manifest itself again in Northwest India in
> the early twenty-first century ... and this is already being reported in the media. (Das Gupta
> et al. 2003, 182)

This kind of argument echoes discussions in the early 1980s, when amniocentesis first began to be used for prenatal sex determination in India (see e.g. Ramanamma and Bambawale 1980). Dharma Kumar suggested that sex-selective abortion was preferable to infanticide or neglect of female children and (whilst she expressed some reservations about overly economistic readings) she predicted that sex selection

> will reduce the supply of women, they will become more valuable, and female children will be
> better cared for and will live longer. We have here a good instrument for balancing the supply of
> and demand for women, and for equating their price all over India (since caste, regional, reli-
> gions and other barriers prevent the movement of women). So in course of time one should
> expect dowries to fall in the North. (Kumar 1983, 63)

Her critics raised several issues. Leela Dube, for example, noted that 'shortages' of women historically had been dealt with by polyandry, abduction, bride purchase, etc. and she predicted that amniocentesis would probably spread as costs of the procedure declined (Dube, 1983a, 1983b). Vishwanath (1983) argued that Kumar's comparison between infanticide and sex-selective abortion ignored the different provenance of the two practices (predominantly rural elites practising infanticide and urban educated upper-class practising amniocentesis) and that she also failed to address the crucial role of hierarchy and hypergamy, which meant that high status men may experience no shortage of women and marry with dowries, whilst poor men may find it hard to marry. My own interventions in this debate also suggested that sex-selection techniques would probably become increasingly popular as norms for lower fertility became more widespread, that economic inequalities mean that marriageable women would be monopolised by wealthy men, that increasing masculinity in the sex ratio was itself symptomatic of women's low value and there was, in any case, no evidence that women are like economic commodities whose value would increase if they were in short supply (Jeffery and Jeffery 1983; Jeffery, Jeffery, and Lyon 1984). In similar vein in response to Bhat and Gupta, Sudha and Rajan (2003) do not consider that shortages of marriageable women will enhance their value on the marriage market: on the contrary, they comment,

> [s]hortfalls in the 'supply' of women will lead to their being subject to greater restrictions, control and violence, as in China, where shortage of marriageable women in some areas has led to kidnapping and sale of women from other regions [Das Gupta and Shuzhuo (sic) 1999]. (4368)

In other words, the relationships between changing marriage payments and demographic change are not as simple as proponents of the 'marriage squeeze' thesis suggest. A closer examination of how marriages are arranged in contemporary India will expand on this point.

Daughter aversion and dowry in the twenty-first century

Multiple marriage markets

Most marriages in India are and will probably continue to be arranged or at least overseen by family elders for many years to come – with some exceptions, such as a select stratum of educated cosmopolitan professionals. Small-scale studies indicate that elders conventionally consider caste/sub-caste, religious community membership, bodily appearance, and the educational level of the potential spouses, their families' relative economic position (with a tendency towards economic isogamy, or hypergamy with the bride 'marrying up'), and the occupations of their members (especially the potential groom) (see e.g. Billig 1991). There is, then, evidence aplenty that the criteria used in selecting suitable marriage partners involve much more than their ages.

Large-scale studies usually enable disaggregation between rural and urban residents down to the district level. Unpacking other intra-regional differences is often not possible, however. Whilst Scheduled Castes and Tribes are routinely detailed separately, other castes tend to be enumerated in combined lists, people's religious community membership is not always registered, and estimating people's economic position is fraught with complexity. Billig (1991), for instance, notes the impossibility of obtaining reliable information on dowry or on sizes of 'marriage pools' disaggregated by caste. Without further disaggregation, though, we can neither understand what is currently happening nor make plausible predictions about the future. Moreover, different settings may require different explanations (Banerjee and Jain 2001), and wider economic processes and class differences – as well as demography – are undoubtedly important at the intra-regional level. Getting to grips with how (if at all) demographic factors impinge on dowry would require more nuanced understandings than the marriage squeeze model has provided so far.

Fertility decline in northern India has been marked by greater discrimination against girls and more widespread use of sex-selective abortion than in the south, so the 'marriage squeeze' might appear to favour potential brides, if not already then in the near future. Once young men's employment prospects are factored in, however, the picture looks very different. Job creation has been an intractable issue throughout India's post-Independence experience. India's apparently spectacular economic performance in recent years has been termed 'jobless growth', with only small numbers of highly trained people entering the newly created jobs in IT and its spin-offs (see e.g. Joshi 2010). The agricultural sector has failed to absorb all the new potential workers, over 90% of workers are in the informal sector, and there is increasing casualisation of employment, with particularly high levels of youth unemployment and widening income inequalities (McNay, Unni, and Cassen 2004). Yet the working-age population will increase by 1.5 times between 2001 and 2026 (McNay, Unni, and Cassen 2004, 170). All these additional people are unlikely to obtain secure or well-paid employment (Acharya, Cassen, and McNay 2004, 206 ff.).

This issue is particularly acute in the large northern states, where the slow decline in fertility (after child mortality had declined substantially) has resulted in fragmentation of land-holdings that reduces both their viability and the rural job opportunities for land-less/land-poor men. This region also has a record of sluggish job creation (Acharya, Cassen, and McNay 2004, 217; Dyson 2010, 40–41). Already, most potentially employed men face considerable problems in the labour market and experience insecure 'informal' sector work and/or extended periods of under- and unemployment or employment in poorly paid activities, even if they are educated (see e.g. Jeffrey, Jeffery, and Jeffery 2008; Jeffrey 2010). Well-settled young men are currently (and will continue to be) in short supply (Banerjee and Jain 2001).

If the marriage squeeze model can provide any insights on daughter aversion and dowry (and I have my doubts), it can only do so if we think in terms of numerous marriage markets (not one marriage market), with multiple marriage squeezes operating at different levels of the class hierarchy simultaneously. One 'marriage squeeze' favours 'suitable boys' who are sought out as grooms. Virtually all young women can marry, but they are siphoned upwards, and women near the top of the hierarchy must provide large dowries. Conversely, a marriage squeeze operates against men without resources or employment prospects who experience a shortage of women (cf. Vishwanath 1983; Billig 1991). Such a class-differentiated scenario has a long history in north-west India. Tried and tested solutions to poor men's compromised marriage chances are not as prestigious as marriage with dowry, but local shortages of brides can be short-circuited by various means: importing brides without dowries – from the northern hills, eastern Gangetic plain, Bangladesh, even Kerala – generally with grooms making some payment (Jeffery, Jeffery, and Lyon 1989, 39–40; Jeffery and Jeffery 1996, 321–244; Das Gupta and Li 1999; Kaur 2004, 2008; Blanchet 2005); de facto fraternal polyandry, when a man's wife is shared with his brothers (Jeffery and Jeffery 1997, 229 ff.); marrying widows to their husband's younger brother (levirate) (e.g. Chowdhry 1994); or marrying a physically challenged woman (see also Dube 1983a, 1983b). In effect, there is a longstanding and continuing 'marriage squeeze' operating against poor men – at the same time as dowry continues to be a sine qua non for marriage higher up the class hierarchy.

Nevertheless, like Bhat, Das Gupta and colleagues (Bhat and Halli 1999; Das Gupta and Li 1999; Das Gupta et al. 2003), some commentators see dowry escalation as a transitory phenomenon that will reverse once more educated grooms are available or when the fertility transition reduces the numbers of potential brides (see e.g. Caldwell, Reddy, and Caldwell 1983; Botticini and Siow 2003). But this 'marriage squeeze' approach rests on a kind of economistic demography that presumes a single perfectly competitive 'market' in potential spouses: 'shortages' or 'surpluses' of grooms or brides affect their relative value as (in effect) economic commodities, which is reflected in marriage payments. Rather than withering, however, dowry is more likely to persist in the upper reaches, whilst increasing numbers of poor men must wait (or 'queue') for many years to marry or even fail to marry at all (cf. Guilmoto 2010, 2012, 81). Given men's poor economic prospects and the marked gender bias in northern India, the 'crisis of masculinity' (Chowdhry 2005) linked to wide class differences in young men's ability to become fully adult (that is, husbands and breadwinners) will be more acute there than in the south.

Gross demographic statistics, then, provide little insight into how marriages are arranged 'on the ground', so using them to make predictions about the long-term impact of contemporary demographic shifts on marriage practices is bound to lead us astray. Moreover, there is a danger that a pre-occupation with demographics, 'marriage squeezes', and how they connect with daughter aversion and dowry distracts attention from other considerations that are crucial to understanding marriage and family life in contemporary India.

Patrilocal marriage and daughter aversion

In India's current 'jobless growth' trajectory, young men face considerable problems in obtaining satisfactory employment. Economic independence is very elusive for women, too, but their 'unemployment' is largely hidden within the home, before and after marriage. Young women, then, generally must comply with their parents' wishes for their marriage, not least because marriage remains their best chance of economic wellbeing. In various other ways, too, compulsory heterosexual marriage reinforces gender inequalities (Basu 2009; Tomalin 2009a). Among these is women's marriage migration, which is a crucial but often ignored issue that has manifold implications.

Marriage in India is generally patrilocal, that is, a bride leaves her parents and joins her husband (and his wider family) (see e.g. Banerjee and Jain 2001; Klasen and Wink 2002; Agrawal and Unisa 2007). Married women in northern India confront several obstacles in maintaining frequent contact with their natal kin: large marriage distances, poor transport and communications, normative restrictions on visits from their natal kin, and controls over their own visits to their natal place. Young women face enormous upheaval: the grief of separation and a disempowering isolation from the supportive networks of their childhood (Jeffery and Jeffery 1988, 1996; Palriwala 1999).[3]

These considerations are exacerbated by dowry. Grooms' families have the upper hand, because of the compulsory character of dowry and the possibility of continuing extortion. Marriage migration in combination with dowry makes young married women vulnerable to dowry demands, violence and harassment, and parents' anxieties about their married daughters might result in their sacrificing the interests of other family members or going into debt.

But women's marriage migration provides strong grounds for not wishing to raise daughters even when there is no dowry system, as Das Gupta et al. (2003) note. In a dowry system with patrilocal residence, a groom's parents make outlays that remain with their household (e.g. providing a home for the young couple) – but they also obtain long-term benefits from the bride, unlike her parents (cf. Sudha and Rajan 2003; Kaur 2008). Overwhelmingly, India's elderly live with relatives, particularly their own adult children, even in the urban areas (see Collard 2000; Croll 2006).

Throughout north India, though, receiving anything from a daughter is disapproved. Thus, Guilmoto (2009) argues, fertility decline has increased the risk of being sonless, not reduced the need for a son. In a low fertility regime, this requirement impacts on the chances that female infants are born or will survive childhood. Women themselves have a significant stake in bearing sons, of course, but they may also be pressurised to bear at least one son or coerced into having a sex-selective abortion. Women's marriage migration in itself, then, perpetuates son preference, because sons and their wives should care for elderly parents. In north India, population-ageing is proceeding more slowly than in the south, and for some decades to come most elderly people will have several sons on whom to rely (see Dyson 2004, 99 ff.). Such support, however, cannot be guaranteed (Kabeer 2000), and, in particular, elderly people without property can exercise little leverage over their sons.

India is now experiencing population-ageing but with little prospect of comprehensive and affordable social provision for the elderly from either the state or the market. We cannot be sure how long the expectation that sons (as distinct from daughters) provide support in old age will persist. Das Gupta (2009) argues that social and economic changes are 'unravelling' the rationale for son preferences, and challenging expectations that daughters migrate to their in-laws' home and that parents should rely only on sons. So far, though, there is only limited evidence for such shifts, mainly in urban areas. In India's north and north-west, where upwards of 70% of the population lives in rural areas, women's marriage migration remains a central part of taken-for-granted assumptions across the class spectrum. Marriage migration is likely to be resilient there for some time to come. Consequently, to the extent that marriage migration is linked to daughter aversion, masculine CSRs are probably not a thing of the past.

Dowry, status competition, and consumerism in neoliberal India

Das Gupta et al. (2003) predict that dowry inflation will gradually reverse because demographic processes are creating a surplus of men and a shortage of brides but there are some further compelling arguments against this optimistic scenario. This requires an examination of dowry in the wider context of the social and economic changes associated most recently with economic liberalisation in India.

That said, though, discussions relating to the early post-1947 period note the relevance of status competition to dowry (e.g. Sharma 1984; Tambiah 1989) and, indeed, dowry has been a concern in areas of northern India that have been characterised by particularly masculine sex ratios since at least the early nineteenth century and that are the main locus of sex-selective abortion today. Aspects of gender politics (such as female infanticide) sometimes featured in nineteenth-century critiques, but commentaries often focused primarily on the indebtedness caused by compulsory displays of ostentatious generosity to honour the groom's family, self-aggrandisement and the staking of status claims by the bride's family, and fear of disgrace (e.g. Ali 1832; Metcalf 1992; Sangari 1999; Sharar 2001). Historians and others link increasing dowry expenditures to the economic changes precipitated during the colonial period: increasing monetisation of the economy; taxation and the resulting indebtedness; markets being flooded with new consumer goods; growing prosperity (for some); and young men employed in the expanding formal economy (in the army, administration, and construction) becoming 'scarce commodities' who could expect large dowries (e.g. Sheel 1999; Oldenburg 2002, 10, 73ff.; Majumdar 2009). For Oldenburg, 'the potential for the custom [dowry] to be converted into blackmail or extortion had increased in an increasingly male-dominated world' (2002, 173). For Srinivas, 'the monster of modern dowry has grown from such humble beginnings' (1984, 4).

Whilst there is a risk of exaggerating the changes precipitated by economic liberalisation, recent data suggest that economic inequality has increased since the early 1990s (Sarkar and Mehta 2010) and India's economic trajectory has been one of remarkable (but 'jobless') economic growth. There are, then, continuing and perhaps worsening 'shortages' of men in secure well-paid jobs. Moreover, hypergamy tends to increase the pool of potential brides available for such men, whilst Srinivasan and Lee (2004) also suggest that low fertility at the top of the economic hierarchy may further enable potential grooms' families to pick and choose. The privileged position of the wealthiest men on the marriage market, then, may fuel dowry escalation, and we should not expect any homogenisation of marriage markets and marriage practices in the near future.

Further, whilst status competition was a crucial element in marriage and dowry long before economic liberalisation, ostentation and display (and thus expense) seem to have taken on a new lease of life since the mid-1980s. Srinivasan (2005, 604ff.) sees an intensification of contradictory processes that have been developing since the late nineteenth century – economic insecurity, casualisation of labour, and widening gaps between rich and poor alongside increased availability of consumer goods and rising aspirations (cf. Srivastava 2002, 259–260; Subramaniam, Remedios, and Mitra 2009). Kaur (2008), Palriwala (2009), and Srinivasan and Lee (2004) all stress the role of dowry and ostentatious marriage ceremonies in people's pursuit of status claims. Even Das Gupta and Li (1999) note that dowry can attain a normative status – which would perhaps operate against the tapering off in dowry inflation that they predict. To this list, Palriwala (2009) adds the identity politics that have beset India since the late 1980s: middle-class Hindu anxiety about loss of identity, she suggests, sparked assertions of primordial identities, including 'traditional' practices associated with weddings and dowry (and, more generally, Hindutva's regressive gender politics).

In addition, the period of economic liberalisation has resulted in the increasing availability of consumer goods, and contemporary dowry is intimately connected to India's increasingly market-driven economy (Narayan 1997, 110; AIDWA 2003). For Basu (2009), dowry became normalised during the 1990s and rose inexorably in the face of consumerism and people's new consumerist identities. People's benchmarks for what is necessary to sustain an acceptable standard of living continually rise (Kaur 2008). Rising aspirations for possessions and domestic comforts, however, coexist with young men's difficulties in attaining economic security. One way of squaring this circle is through dowry: as Palriwala (1989) put it, women are commodities themselves but also the conduits through which commodities are obtained in a largely patrilineal and patrilocal society. For 'many in the middle and lower middle class, dowry is the means to acquire desired consumer goods, capital for investment, bribes to "buy" secure jobs, or an investment which may draw in further wealth' (Palriwala 2009, 161; see also Srinivasan and Lee 2004). Faced with growing uncertainties about their livelihood prospects, many young men (and their elders) are increasingly unwilling to arrange marriages without a cash incentive: some men can access consumer goods and cash only via their wife's dowry, whilst particular occupations and castes may have specified rate charts (Banerjee and Jain 2001). Some men with good employment prospects might opt for marriages without dowries, but (as Kaur 2007 notes) they are in a position to command larger dowries than other men. And, since there is no sign that the desire for material goods is abating, this aspect of dowry is also likely to persist.

In combination with women's marriage migration and the growth of consumerism, moreover, the expectation that the bride's family will continually provide goods and cash throughout the life of the marriage can place the wife in a vulnerable position – one that puts further pressure on her parents to ensure that they are as generous as possible (cf. Narayan 1997, 83–117). Demands are a feature not only of the period prior to the marriage or of the dowry narrowly understood and parents and married women alike express concerns about security and protection from ill-treatment in the marital home (Basu 2001, Chapter 3; Srinivasan and Lee 2004; Palriwala 2009). As Bossen (1988, 141) starkly noted: 'In effect, there is a risk that a woman can become a hostage to a family that has already fully claimed the ransom from the bride's parents.'

Demography and political complacency

I am, then, not at all sanguine that declining fertility in conjunction with masculine CSRs will (in due course) result in the reversal of a marriage squeeze that supposedly leads to a decline in dowry (because young men would no longer be in short supply) and a gradual reduction in daughter aversion (which would be evidenced in a declining incidence of sex-selective abortion).

Whilst demographic processes may play some part in dowry escalation and daughter aversion, the situation is much more complex than this kind of demographic determinism allows – and thus the prognosis for the future is also more complicated and uncertain. Indeed, there are compelling reasons for thinking that dowry and daughter aversion can both continue, even if demographic shifts might appear to suggest the reverse. Given marriage migration and the lack of social provision for the care of the elderly (on the one hand) and status competition and consumerism (on the other), daughter aversion is likely to continue and brides' parents will still provide as much dowry as possible. People in the middle ranks will mimic those above them in their efforts to make status claims, and many people will make outlays that they can ill-afford. Meanwhile, men at the bottom of the class hierarchy may be unable to contract marriages that bring a dowry and may have to queue for years to marry at all.

In other words, dowry and daughter aversion should be analysed in tandem with competitive display and rising expectations, and with India's changing livelihood options and how they play out in gendered ways at different levels in the class hierarchy. They are not best understood in terms of timeless or archaic cultural traditions, religious texts or emotion, or even as something that is significant for gender politics narrowly understood (Narayan 1997, 83–117; Basu 2005b; Tomalin 2009b). Nor should they be too closely coupled to the lens of 'marriage squeeze' and demographic change.

My objections here are not only theoretical, however. A common presumption among demographers (and others) is that declining fertility benefits women. Some benefits no doubt do result – for instance in reducing women's life-time risk of maternal mortality and maternal depletion. Yet fertility decline in India, especially in the north, has been gender-biased. How can that be read in a wholly positive light? Why should we accept that it might lead to a reversal of the marriage squeeze against women and a reduction in dowry that benefits women? The belief in the capacity of demographic change to enhance women's position is paradoxical and oversimplifies an extremely complex situation. And such misplaced optimism that demographic shifts will result in the erosion of dowry and daughter aversion runs the danger of allowing space for political complacency about the trajectory of gender politics in neoliberal India.

Since the early 1980s, feminist activism in India has campaigned on numerous issues, among them dowry and sex-selective abortion (see e.g. AIDWA 2003; Basu 2005a; Bradley, Tomalin, and Subramaniam 2009). Yet, as Harriss-White (1999) argues, declines in fertility coupled with the increasing need to provide old-age support – a form of 'demographic "structural adjustment" without precedent' (148–149) – pull in the opposite direction from the Indian government's piecemeal (and ineffective) efforts to eradicate dowry and sex-selective abortion. Many feminists, indeed, would argue that focusing on specific problems such as dowry and sex-selective abortion merely tackles symptoms rather than their deeper causes. Dowry and daughter aversion are embedded in gender politics more generally and this implies, as Tomalin (2009a) puts it, profound structural changes that transform gender relations rather than mere window-dressing. These would include, for instance, addressing women's property rights and access to education and employment. But gender politics themselves are embedded in the rapidly changing economic and political context of a competitive, highly unequal and increasingly consumerist society. Jobless growth adversely affects young men's prospects, but it also fails to provide much scope for women's economic independence. This all implies, then, a rather daunting and long-term political agenda focusing not solely on the politics of the domestic sphere but also on the pernicious effects of national and global economic processes – and certainly not in-activism bred of complacent assumptions that things will all work out for the best as a result of inexorable demographic processes.

Acknowledgements

The sole responsibility for what I have written here is mine. I am grateful to colleagues for their comments on earlier versions of this paper, and especially to Minna Säävälä, Sirpa Tenhunen and two anonymous reviewers. This paper was made possible by a British Academy/Leverhulme Trust Senior Research Fellowship and a Leverhulme Trust Research Fellowship (2009–2010). I am grateful for their support.

Notes

1. It is important to note here that quantifying the potential 'surplus' of marriageable men depends on the parameters used and other demographers have arrived at different conclusions on this score. See, for instance, Neelakantan and Tertilt (2008) as well as Guilmoto (2012).
2. Demographers of India usually express sex ratios such as SRB and CSR in terms of the number of females per 1000 males, whilst the international convention is for sex ratios to express the number of males in relation to females.
3. Jeffrey and Doron (2013), however, indicate how mobile phones may be moderating this.

References

Acharya, Shankar, Robert Cassen, and Kirsty McNay. 2004. "The Economy – Past and Future." In *Twenty-First Century India: Population, Economy, Human Development, and the Environment*, edited by T. Dyson, R. Cassen, and L. Visaria, 202–227. New Delhi: Oxford University Press.
Agnihotri, Satish Balram. 2000. *Sex Ratio Patterns in the Indian Population: A Fresh Exploration*. New Delhi: Sage.
Agnihotri, Satish Balram. 2001a. "Declining Infant and Child Mortality in India." *Economic and Political Weekly* 36 (3): 228–33.
Agnihotri, Satish Balram. 2001b. "Rising Sons and Setting Daughters: Provisional Results from the 2001 Census." In *Enduring Conundrum: India's Sex Ratio* (Essays in honour of Asok Mitra), edited by V. Mazumdar and N. Krishnaji, 199–201. Delhi: Rainbow Publishers (for Centre for Women's Development Studies).
Agnihotri, Satish Balram. 2003. "Survival of the Girl Child: Tunnelling out of the Chakravyuha." *Economic and Political Weekly* 38 (41): 4351–4360.
Agrawal, Sutapa, and Sayeed Unisa. 2007. "Discrimination from Conception to Childhood: A Study of Girl Children in Rural Haryana, India." In *Watering the Neighbour's Garden: The Growing*

Demographic Female Deficit in Asia, edited by I. Attané and C. Z. Guilmoto, 247–266. Paris: Committee for International Cooperation in National Research in Demography (CICRED).

AIDWA (All India Democratic Women's Association), ed. 2003. *Expanding Dimensions of Dowry*. New Delhi: All India Democratic Women's Association.

Ali, Mrs. Meer Hasan. 1832. *Observations on the Mussulmauns of India: Descriptive of Their Manners, Customs, Habits and Religious Opinions*. 2 vols. London: Parbury, Allen (reprinted in 1973 by Idarah-i Adabiyat-i Delli, Delhi).

Arnold, Fred, Sunita Kishor, and T. K. Roy. 2002. "Sex Selective Abortions in India." *Population and Development Review* 28 (4): 759–785.

Arokiasamy, Perianayagam. 2004. "Regional Patterns of Sex Bias and Excess Female Child Mortality." *Population* 59 (6): 831–863.

Arokiasamy, Perianayagam. 2007. "Sex Ratio at Birth and Excess Female Child Mortality in India: Trends, Differentials and Regional Patterns." In *Watering the Neighbour's Garden: The Growing Demographic Female Deficit in Asia*, edited by I. Attané and C. Z. Guilmoto, 49–72. Paris: Committee for International Cooperation in National Research in Demography (CICRED).

Arokiasamy, Perianayagam, and Abhishek Gautam. 2008. "Neonatal Mortality in the Empowered Action Group States of India: Trends and Determinants." *Journal of Biosocial Science* 40 (2): 183–201.

Attané, Isabelle, and Christophe Z. Guilmoto. 2007. "Introduction." In *Watering the Neighbour's Garden: The Growing Demographic Female Deficit in Asia*, edited by I. Attané and C. Z. Guilmoto, 1–22. Paris: Committee for International Cooperation in National Research in Demography (CICRED).

Aziz, Abdul. 1983. (No title). *Economic and Political Weekly* 18: 603–604.

Banerjee, Nirmala, and Devaki Jain. 2001. "Indian Sex Ratios Through Time and Space: Development from Women's Perspective." In *Enduring Conundrum: India's Sex Ratio* (Essays in honour of Asok Mitra), edited by V. Mazumdar and N. Krishnaji, 73–119. Delhi: Rainbow Publishers (for Centre for Women's Development Studies).

Basu, Srimati. 2001. *She Comes to Take Her Rights: Indian Women, Property, and Propriety*. New Delhi: Kali for Women.

Basu, Srimati, ed. 2005a. *Dowry and Inheritance*. New Delhi: Women Unlimited.

Basu, Srimati. 2005b. "The Politics of Giving: Dowry and Inheritance as Feminist Issues." In *Dowry and Inheritance*, edited by S. Basu, i-liv. New Delhi: Women Unlimited (an associate of Kali for Women).

Basu, Srimati. 2009. "Legacies of the Dowry Prohibition Act in India: Marriage Practices and Feminist Discourses." In *Dowry: Bridging the Gap Between Theory and Practice*, edited by T. Bradley, E. Tomalin, and M. Subramaniam, 177–196. New Delhi: Women Unlimited.

Bhalotra, Sonia, Christine Valente, and Arthur van Soest. 2010a. "The Puzzle of Muslim Advantage in Child Survival in India." *Journal of Health Economics* 29 (2): 191–204.

Bhalotra, Sonia, Christine Valente, and Arthur van Soest. 2010b. "Religion and Childhood Death in India." In *Handbook of Muslims in India*, edited by A. Shariff and R. Basant, 123–164. New Delhi: Oxford University Press.

Bhat, P. N. Mari. 2002a. "On the Trail of 'Missing' Indian Females I: Search for Clues." *Economic and Political Weekly* 37 (51): 5105–18.

Bhat, P. N. Mari. 2002b. "On the Trail of 'Missing' Indian Females II: Illusion and Reality." *Economic and Political Weekly* 37 (51): 5244–5263.

Bhat, P. N. Mari, and A. J. Francis Zavier. 2003. "Fertility Decline and Gender Bias in Northern India." *Demography* 40 (4): 637–657.

Bhat, P. N. Mari, and A. J. Francis Zavier. 2007. "Factors Influencing the Use of Prenatal Diagnostic Techniques and Sex Ratio at Birth in India." In *Watering the Neighbour's Garden: The Growing Demographic Female Deficit in Asia*, edited by I. Attané and C. Z. Guilmoto, 131–160. Paris: Committee for International Cooperation in National Research in Demography (CICRED).

Bhat, P. N. Mari, and Shiva S. Halli. 1999. "Demography of Brideprice and Dowry: Causes and Consequences of the Indian Marriage Squeeze." *Population Studies* 53 (2): 129–148.

Billig, Michael S. 1991. "The Marriage Squeeze on High-Caste Rajasthani Women." *Journal of Asian Studies* 50 (2): 341–360.

Blanchet, Thérèse. 2005. "Bangladeshi Girls Sold as Wives in North India." *Indian Journal of Gender Studies* 12 (2&3): 305–334.

Bossen, Laurel. 1988. "Toward a Theory of Marriage: The Economic Anthropology of Marriage Transactions." *Ethnology* 27 (2): 127–144.

Botticini, Maristella, and Aloysius Siow. 2003. "Why Dowries?" *American Economic Review* 93 (4): 1385–1398.

Bradley, Tamsin, Emma Tomalin, and Mangala Subramaniam, eds. 2009. *Dowry: Bridging the Gap Between Theory and Practice*. New Delhi: Women Unlimited.

Caldwell, J. C., P. H. Reddy, and Pat Caldwell. 1983. "The Causes of Marriage Change in South India." *Population Studies* 37 (3): 343–361.

Chowdhry, Prem. 1994. *The Veiled Women: Shifting Gender Equations in Rural Haryana 1880–1990*. Delhi: Oxford University Press.

Chowdhry, Prem. 2005. "Crisis of Masculinity in Haryana: The Unmarried, the Unemployed and the Aged." *Economic and Political Weekly* 40 (49): 5189–5198.

Collard, David. 2000. "Generational Transfers and the Generational Bargain." *Journal of International Development* 12 (4): 342–351.

Croll, Elisabeth J. 2006. "The Intergenerational Contract in the Changing Asian Family." *Oxford Development Studies* 34 (4): 473–491.

Das Gupta, Monica. 2009. *Family Systems, Political Systems, and Asia's 'Missing Girls': The Construction of Son Preference and its Unraveling*. Policy Research Working Paper 5148. World Bank.

Das Gupta, Monica, Zhenghua Jiang, Bohua Li, Zhenming Xie, Chung Woojin, and Hwa-Ok Bae. 2003. "Why Is Son Preference So Persistent in East and South Asia? A Cross-country Study of China, India and the Republic of Korea." *Journal of Development Studies* 40 (2): 153–187.

Das Gupta, Monica, and P. N. Mari Bhat. 1997. "Fertility Decline and Increased Manifestation of Sex Bias in India." *Population Studies* 51 (3): 307–315.

Das Gupta, Monica, and Li Shuzhuo. 1999. "Gender Bias in China, South Korea and India 1920–1990: Effects of War, Famine and Fertility Decline." *Development and Change* 30 (3): 619–652.

Dube, Leela. 1983a. "Amniocentesis Debate Continued." *Economic and Political Weekly* 18 (38): 1633–1635.

Dube, Leela. 1983b. "Misadventures in Amniocentesis." *Economic and Political Weekly* 18 (8): 279–280.

Dyson, Tim. 2001. "The Preliminary Demography of the 2001 Census of India." *Population and Development Review* 27 (2): 341–356.

Dyson, Tim. 2004. "India's Population – the Future." In *Twenty-First Century India: Population, Economy, Human Development and the Environment*, edited by T. Dyson, R. Cassen, and L. Visaria, 74–107. New Delhi: Oxford University Press.

Dyson, Tim. 2010. "Growing Regional Variation: Demographic Change and Its Implications." In *Change and Diversity: Economics, Politics and Society in Contemporary India* (*Proceedings of the British Academy*, 159), edited by A. Heath and R. Jeffery, 19–46. Oxford: Oxford University Press.

Goodkind, Daniel. 1996. "On Substituting Sex Preference Strategies in East Asia: Does Prenatal Sex Selection Reduce Postnatal Discrimination?" *Population and Development Review* 22 (1): 111–125.

Guilmoto, Christophe Z. 2008. "Economic, Social and Spatial Dimensions of India's Excess Child Masculinity." *Population-E* (English) 63 (1): 91–118.

Guilmoto, Christophe Z. 2009. "The Sex Ratio Transition in Asia." *Population and Development Review* 35 (3): 519–549.

Guilmoto, Christophe Z. 2010. "Longer-Term Disruptions to Demographic Structures in China and India Resulting from Skewed Sex Ratios at Birth." *Asian Population Studies* 6 (1): 3–24.

Guilmoto, Christophe Z. 2012. "Skewed Sex Ratios at Birth and Future Marriage Squeeze in China and India, 2005–2100." *Demography* 49 (1): 77–100.

Guilmoto, Christophe Z., and Isabelle Attané. 2007. "The Geography of Deteriorating Child Sex Ratio in China and India." In *Watering the Neighbour's Garden: The Growing Demographic Female Deficit in Asia*, edited by I. Attané and C. Z. Guilmoto, 109–129. Paris: Committee for International Cooperation in National Research in Demography (CICRED).

Harriss-White, Barbara. 1999. "Gender-Cleansing: The Paradox of Development and Deteriorating Female Life Chances in Tamil Nadu." In *Signposts: Gender Issues in Post-Independence India*, edited by R. Sunder Rajan, 125–154. New Delhi: Kali for Women.

ICSSR (Indian Council of Social Science Research). 1975. *Status of Women in India: Synopsis of the Report of the National Committee on the Status of Women (1971–74)*. Indian Council of Social Science Research.

Jeffery, Patricia, and Roger Jeffery. 1996. *Don't Marry Me to a Plowman! Women's Everyday Lives in Rural North India*. Boulder, CO: Westview Press and New Delhi: Vistaar.

Jeffery, Patricia, Roger Jeffery, and Andrew Lyon. 1988. "When Did You Last See Your Mother? Aspects of Female Autonomy in Rural North India." In *Micro-approaches to Demographic Research*, edited by J. Caldwell, A. Hill, and V. Hull, 321–333. London: Kogan Page International.

Jeffery, Patricia, Roger Jeffery, and Andrew Lyon. 1989. *Labour Pains and Labour Power: Women and Childbearing in India*. London: Zed Books.

Jeffery, Roger, and Patricia Jeffery. 1983. "Female Infanticide and Amniocentesis." *Economic and Political Weekly* 18 (16–17): 654–656.

Jeffery, Roger, and Patricia Jeffery. 1997. *Population, Gender and Politics: Demographic Change in Rural North India*. Cambridge: Cambridge University Press.

Jeffery, Roger, Patricia Jeffery, and Andrew Lyon. 1984. "Female Infanticide and Amniocentesis." *Social Science and Medicine* 19 (11): 1207–1212.

Jeffrey, Craig. 2010. *Timepass: Youth, Class, and the Politics of Waiting in India*. Stanford: Stanford University Press.

Jeffrey, Craig, Patricia Jeffery, and Roger Jeffery. 2008. *Degrees Without Freedom? Education, Masculinities and Unemployment in North India*. Stanford: Stanford University Press.

Jeffrey, Robin, and Assa Doron. 2013. *The Great Indian Phone Book: How Cheap Mobile Phones Change Business, Politics and Daily Life*. London: Hurst & Co.

John, Mary E., Ravinder Kaur, Rajni Palriwala, Saraswati Raju, and Alpana Sagar. 2008. *Planning Families, Planning Gender: The Adverse Child Sex Ratio in Selected Districts of Madhya Pradesh, Rajasthan, Himachal Pradesh, Haryana and Punjab*. ActionAid India and International Development Research Centre Canada. Accessed February 18, 2010. http://www.cwds.ac.in/PlanningFamiliesPlanningGender.pdf

Joshi, Vijay. 2010. "Economic Resurgence, Lopsided Reform and Jobless Growth." In *Diversity and Change in Modern India: Economic, Social and Political Approaches*, edited by A. Heath and R. Jeffery, 73–106. Oxford: Oxford University Press.

Kabeer, Naila. 2000. "Inter-generational Contracts, Demographic Transitions and the 'Quantity-Quality' Tradeoff: Parents, Children and Investing in the Future." *Journal of International Development* 12 (4): 463–482.

Kaur, Ravinder. 2004. "Across-Region Marriages: Poverty, Female Migration and the Sex Ratio." *Economic and Political Weekly* 39 (25): 2595–2603.

Kaur, Ravinder. 2007. "Declining Juvenile Sex Ratios: Economy, Society and Technology Explanations from Field Evidence." *Margin – The Journal of Applied Economic Research* 1 (1): 231–245.

Kaur, Ravinder. 2008. "Dispensable Daughters and Bachelor Sons: Sex Discrimination in North India." *Economic and Political Weekly* 43 (30): 109–114.

Klasen, Stephan, and Claudia Wink. 2002. "A Turning Point in Gender Bias in Mortality? An Update on the Number of Missing Women." *Population and Development Review* 28 (2): 285–312.

Kulkarni, Purushottam. 2012. "India's Child Sex Ratio: Worsening Imbalance." *Indian Journal of Medical Ethics* 9 (2): 112–114.

Kumar, Dharma. 1983. "Amniocentesis Again." *Economic and Political Weekly* 18 (24): 1075–1076.

Kumari, Ranjana. 1989. *Brides Are Not for Burning: Dowry Victims in India*. London: Sangam Books.

Larsen, Mattias. 2011. *Vulnerable Daughters in India: Culture, Development and Changing Contexts*. New Delhi: Routledge.

Majumdar, Rochana. 2009. *Marriage and Modernity: Family Values in Colonial Bengal*. Durham: Duke University Press.

Mayer, Peter. 1999. "India's Falling Sex Ratios." *Population and Development Review* 25 (2): 323–343.

McNay, Kirsty, Jeemol Unni, and Robert Cassen. 2004. "Employment." In *Twenty-First Century India: Population, Economy, Human Development and the Environment*, edited by T. Dyson, R. Cassen, and L. Visaria, 158–177. New Delhi: Oxford University Press.

Metcalf, Barbara Daly. 1992. *Perfecting Women: Maulana Ashraf 'Ali Thanawi's Bihishti Zewar: A Translation with Commentary*. Delhi: Oxford University Press.

Narayan, Uma. 1997. *Dislocating Cultures: Identities, Traditions, and Third World Feminism.* New York: Routledge.

Neelakantan, Urvi, and Michèle Tertilt. 2008. "A Note on Marriage Market Clearing." *Economics Letters* 101 (2): 103–105.

Oldenburg, Philip. 1992. "Sex Ratio, Son Preference and Violence in India: A Research Note." *Economic and Political Weekly* 27 (49–50): 2657–2662.

Oldenburg, Veena Talwar. 2002. *Dowry Murder: The Imperial Origins of a Cultural Crime.* New York: Oxford University Press.

Palriwala, Rajni. 1989. "Reaffirming the Anti-Dowry Struggle." *Economic and Political Weekly* 24 (17): 942–944.

Palriwala, Rajni. 1999. "Transitory Residents, Invisible Workers: Rethinking Locality and Incorporation in a Rajasthan Village." In *From Myth to Markets: Essays on Gender*, edited by K. Sangari and U. Chakravarti, 237–273. New Delhi: Manohar (for Indian Institute of Advanced Study, Shimla).

Palriwala, Rajni. 2009. "The Spider's Web: Seeing Dowry, Fighting Dowry." In *Dowry: Bridging the Gap Between Theory and Practice*, edited by T. Bradley, E. Tomalin, and M. Subramaniam, 144–176. New Delhi: Women Unlimited.

Patel, Tulsi, ed. 2007. *Sex-Selective Abortion in India: Gender, Society and New Reproductive Technologies.* New Delhi: Sage Publications.

Purewal, Navtej K. 2010. *Son Preference: Sex Selection, Gender and Culture in South Asia.* Oxford: Berg.

Ramanamma, A., and Usha Bambawale. 1980. "The Mania for Sons: An Analysis of Social Values in South Asia." *Social Science and Medicine* 14B (2): 107–110.

Sambrani, Rita Bhandari, and Shreekant Sambrani. 1983. "Economics of Bride-Price and Dowry." *Economic and Political Weekly* 18 (36–37): 601–603.

Sangari, Kumkum. 1999. *Politics of the Possible: Essays on Gender, History, Narrative, Colonial English.* New Delhi: Tulika.

Sarkar, Sandip, and Balwant Singh Mehta. 2010. "Income Inequality in India: Pre- and Post-reform Periods." *Economic and Political Weekly* 45 (37): 45–55.

Sharar, Abdul Halim. 2001. *Lucknow: The Last Phase of an Oriental Culture.* Translated and edited by E. S. Harcourt and Fakhir Hussain. New Delhi: Oxford University Press.

Sharma, Ursula. 1984. "Dowry in North India: Its Consequences for Women." In *Women and Property, Women as Property*, edited by R. Hirschon, 62–74. London: Croom Helm.

Sheel, Ranjana. 1999. *The Political Economy of Dowry: Institutionalization and Expansion in North India.* New Delhi: Manohar.

Siddhanta, Suddhasil, Satish B. Agnihotri, and Debasish Nandy. 2009. "Sex Ratio Patterns among the Scheduled Castes in India 1981–2001." Paper presented at World Congress of the International Union for the Scientific Study of Population, in Marrakech, Morocco, September 27–October 2.

Srinivas, M. N. 1984. *Some Reflections on Dowry.* New Delhi: Oxford University Press.

Srinivasan, Padma, and Gary R. Lee. 2004. "The Dowry System in Northern India: Women's Attitudes and Social Change." *Journal of Marriage and Family* 66 (5): 1108–1117.

Srinivasan, Sharada. 2005. "Daughters or Dowries? The Changing Nature of Dowry Practices in South India." *World Development* 33 (4): 593–615.

Srivastava, Nisha. 2002. "Multiple Dimensions of Violence Against Rural Women in Uttar Pradesh: Macro and Micro Realities." In *The Violence of Development: The Politics of Identity, Gender and Social Inequalities in India*, edited by K. Kapadia, 235–291. New Delhi: Kali for Women.

Subramaniam, Mangala, Karen Remedios, and Debarashmi Mitra. 2009. "Dowry and Transnational Activism." In *Dowry: Bridging the Gap Between Theory and Practice*, edited by T. Bradley, E. Tomalin, and M. Subramaniam, 197–225. New Delhi: Women Unlimited.

Sudha, S., and S. Irudaya Rajan. 1999. "Female Demographic Disadvantage in India 1981–1991: Sex Selective Abortions and Female Infanticide." *Development and Change* 30 (3): 585–618.

Sudha, S., and S. Irudaya Rajan. 2003. "Persistent Daughter Disadvantage: What Do Estimated Sex Ratios at Birth and Sex Ratios of Child Mortality Risk Reveal?" *Economic and Political Weekly* 38 (41): 4361–4369.

Tambiah, Stanley J. 1989. "Bridewealth and Dowry Revisited: The Position of Women in Sub-Saharan Africa and North India." *Current Anthropology* 30 (4): 413–435.

Tomalin, Emma. 2009a. "Conclusion." In *Dowry: Bridging the Gap Between Theory and Practice*, edited by T. Bradley, E. Tomalin, and M. Subramaniam, 226–241. New Delhi: Women Unlimited.

Tomalin, Emma. 2009b. "Introduction." In *Dowry: Bridging the Gap Between Theory and Practice*, edited by T. Bradley, E. Tomalin, and M. Subramaniam, 1–28. New Delhi: Women Unlimited.

Vatuk, Sylvia. 2007. "The 'Cancer of Dowry' in Indian Muslim Marriages: Themes in Popular Rhetoric from the South Indian Muslim Press." In *Living with Secularism: The Destiny of India's Muslims*, edited by M. Hasan, 155–175. New Delhi: Manohar.

Visaria, Leela. 2004a. "The Continuing Fertility Transition." In *Twenty-First Century India: Population, Economy, Human Development and the Environment*, edited by T. Dyson, R. Cassen, and L. Visaria, 57–73. New Delhi: Oxford University Press.

Visaria, Leela. 2004b. "Mortality Trends and the Health Transition." In *Twenty-First Century India: Population, Economy, Human Development and the Environment*, edited by T. Dyson, R. Cassen, and L. Visaria, 32–56. New Delhi: Oxford University Press.

Vishwanath, L. S. 1983. "Misadventures in Amniocentesis." *Economic and Political Weekly* 18 (11): 406–407.

Domestic violence made public: a case study of the use of alternative dispute resolution among underprivileged women in Bangladesh

Laila Ashrafun[a,b] and Minna Säävälä[a,c]

[a]Department of Social Research, University of Helsinki, Helsinki, Finland; [b]Department of Sociology, Shahjalal University of Science & Technology, Kumargaon, Bangladesh; [c]Population Research Institute, Väestöliitto, Helsinki, Finland

Disputes within the family and cases of domestic violence in Bangladesh have traditionally been resolved either among kinsmen or in village *panchayats* where young women have had less opportunity to be heard. This article describes a third-sector initiative in alternative dispute resolution (ADR) in cases of domestic violence in Sylhet, Bangladesh. Data from 20 case studies of female victims of domestic violence, interviews with six lawyers, and participant observation in 10 mediations in a counselling centre are used to examine how this kind of social innovation affects the bargaining position of battered women in a kin-based, patrilineal and patrilocal society. The battered women involved in these mediations were rural or urban, less educated, and economically disadvantaged. The findings illustrate the limitations of ADR in cases of domestic violence. ADR as practiced in Sylhet, Bangladesh provides poor women a chance to be publicly heard in mediations of their domestic crises. However, ADR often fails to deliver lasting, just, and socially progressive solutions. The adoption of ADR practices should not be considered as an alternative to the development of the formal judicial system, because it lacks the power to enforce agreements and supports the hegemonic status quo, leaving the battered woman and her natal family with very limited options.

Introduction

As in other parts of South Asia, in Bangladesh gender-based violence has been recognized as one of the most blatant manifestations of gender asymmetry (UNICEF 2011). Estimates of the prevalence of married women's experiences of domestic violence in a lifetime range between 32% (Steele, Amin, and Naved 1998) and 72% (BIDS 2004).[1] Domestic violence is aggravated by poverty, poor living conditions, and lack of resources. A recent study found that the prevalence of reported past-year physical spousal violence in Dhaka, Bangladesh is higher in slums (35%) than in non-slums (20%) (Sambisa et al. 2011).

The problem of domestic violence against women has been acknowledged by international and national developmental agencies as well as the government of Bangladesh. Many organizations[2] and the Bangladeshi Ministry of Social Welfare are working to combat domestic violence (Ameen 2005; Chowdhury 2011; Jahan 1994). Third-sector organizations have tried to stop the trafficking in women and children and they provide legal counselling and training in community awareness against domestic violence (Chowdhury 2011).

The judicial system still does not render justice to underprivileged female victims of domestic violence in Bangladesh (Ashrafun 2013). Resolving cases of domestic violence has been assigned either to the community level via kin groups or local communities or to the formal legal arena governed by state institutions. The legal process is often expensive, slow, and unpredictable and consequently does not work well in cases of family disputes, whereas the local community may be heavily biased towards the interests of the power holders, i.e. elders and men.

This article examines how women in socially vulnerable groups make use of an alternative dispute resolution (ADR) process in a collaborative project between a judicial body and a third-sector agent in Sylhet, Bangladesh. By presenting ethnographic and interview data on cases of domestic violence against women of the lower socio-economic strata and on mediation sessions, we will seek an understanding of how informal mediation affects the bargaining position of the battered women. Can such socially innovative mediation alleviate their situation and does it lead to sustainable solutions? Do women get a chance to express their views in the legal arena and obtain justice to establish their rights? We argue that ADR in the form that it is practiced at this particular counselling centre actually leaves the women in a very vulnerable position.

ADR was developed in the USA in the 1970s and 1980s, particularly as a court reform project (Cohen 2006; Nader and Grande 2002) to provide an alternative way of reconciling civil disagreements. ADR refers to dispute resolution mechanisms such as negotiation, mediation, arbitration, conciliation, early neutral evaluation, and mini-trial mediation (Cohen 2006). It is a social innovation developed for resolving disputes with the aim of enabling disagreeing parties to reach agreement without lengthy and costly litigation. The US-style ADR includes the involvement of a neutral third party, a mediator, who facilitates the resolution process, although she or he does not impose a resolution on the parties.

ADR has travelled from the USA to other parts of the world and has been adopted as a common component of development projects intended to modernize and rationalize state judicial systems, for example, by USAID and the World Bank (Capulong 2012; Cohen 2006; Maru 2010). Nader and Grande (2002) view the effects of the US-based ADR as so penetrating that they speak of a 'global ADR revolution' in the justice field. When the original model has been adopted in vastly different socio-economic and developmental contexts, it has, however, not remained unchanged: 'Mediation changes as it travels; its instantiation anywhere is subject to local variation and invention as it makes contact with state and customary law, politics, and social struggles' (Cohen 2006; see also Nader and Grande 2002). Here, we will examine the practical application of ADR when adapted to a kin-based Bangladeshi society in the context of domestic violence.

Although ADR as a distinct social innovation was developed in the USA only a few decades ago, forms of community mediation as an informal social process of delivering justice has long historical roots. It has been present in some form in all societies prior to the introduction of formal legal institutions. In South Asia, village elders or *panchayats* have traditionally acted as mediators or arbiters in the place of formal legal bodies (Sharma 2004). Informal mediation, *shalish,* is a very common practice in Bangladesh. A *shalish* committee typically consists of elderly, honourable and influential members of a community; thus, there is no neutral arbiter or mediator in the case, unlike in the ADR systems. The resolution is made from the perspective of community interest, not the individual interest of the parties involved, unlike in the US-type ADR. To enforce its resolution, a *shalish* committee uses monetary fines and social boycotting of the offender's family, which are both absent in ADR. Basically, in traditional *shalish* processes, women are not allowed to be present during the arbitration even if their own case is under scrutiny. They are usually represented by their male relatives. Therefore, this alternative justice system may favour men over women (Jahan 2005) and the interests of the community power holders.

Introducing the US-style ADR in South Asia has been an innovation in the sense that it occupies a space between the judiciary and civil society. In India, ADR has been an institutionalized part of the legal system since the Arbitration and Conciliation Act of 1996 and even prior to that, based on other acts (Sharma 2004). In Bangladesh, where the formal legal system has been built on the same colonial legacy as in India, ADR does not yet have a similar level of institutionalization. Following the model provided by the Canadian, American, and Indian judicial systems, ADR has been tried out in some Bangladeshi courts as an alternative to formal legal procedures (Kamal 2004).

The global exportation of ADR has been based on its image as an apolitical and neutral social innovation that can benefit citizens in conditions of poorly functioning state-level judicial systems. Since the 1990s, a growing number of scholars have raised their voice to oppose the universal advocacy of US-style ADR on grounds of social justice, arguing that mediation may end up serving the interests of legal and social elites and diminishing opportunities for social change (e.g. Cohen 2006; Grillo 1991; Nader and Grande 2002).

Data and methodology

The data in this article derive from a wider qualitative study on domestic violence in Sylhet, Bangladesh (Ashrafun 2013). Here we have examined in detail the cases of 20 poor young female victims at a Bangladesh Legal Aid and Services Trust (BLAST) counselling centre in the city of Sylhet. The aim is to provide an in-depth representation of the family life of and domestic violence perpetrated against young women by their affines[3] and the effectiveness of *shalish* as a form of dispute resolution. The women in the case studies are young wives aged between 15 and 28 years. Apart from a couple of women who had attended school till the 9th grade, the women have very little or no education and are housewives or labourers. The most common perpetrator of violence is the husband, but there are also cases of the women's mother-in-law, father-in-law, or other affine being the aggressor.

The district court in Sylhet has adopted a cooperative initiative between the court and an NGO, the BLAST, to mediate in family disputes such as cases of domestic violence and divorce. BLAST is a prominent legal services organization in Bangladesh, currently operating in 19 districts across the country. BLAST's mission is to make the legal system accessible to the poor and marginalized by providing free services. The majority of those who come to BLAST prefer mediation to litigation due to the reduced cost and time involved. If the parties fail to settle the dispute through mediation, the advocates of the counselling centre take the case to the court and represent the client free of charge.

Through informal mediation, the NGO tries to give the underprivileged women an opportunity to have their voices heard with the support of their natal family members, kin, acquaintances, and advocates. As a result of the mediation, the husband and wife are either reunited, the parties sign a bond to abide by an agreement, or a divorce takes place along with a settlement for the maintenance and custody of children.

The data consist of interviews with 20 women with a mediation case as well as six advocates, observation of 10 mediations, and informal talks with other participants in the mediations. The case study interviews with the women, whose cases of domestic violence and arbitration this article analyses, were conducted in privacy to maintain confidentiality. No tape recorder was used in these discussions with the female victims due to the emotional crisis which required a particularly sensitive and personal environment (Liamputtong 2007). Copious notes were taken and transcribed in greater detail following the interviews with the women. The interviewees were informed about the study and its objectives and their voluntary oral consent was secured prior to their inclusion in the study. The interviews were carried out in the local Bengali language by a married female national, Laila Ashrafun. Conversations with the women about their problems were lengthy, lasting three hours or more.

With the cooperation of the advocates of BLAST, Laila Ashrafun observed 10 mediations. The participants in the mediations were asked for their consent for the researcher's presence. The coordinator of the counselling centre heading the sessions explained the role of the researcher and sought the participants' consent for her presence and the use of the *shalish* as research data. In none of the mediations did the clients refuse the researcher's presence or object to the use of their mediation as research data. Individual consent was sought from the guardians of the young female victims (usually a father or mother) and from the main negotiators of the offenders' party. There were typically numerous participants, from 6 to 12 people, in each mediation session.

For the sake of confidentiality, all personal data that could lead to recognition is omitted or transformed and pseudonyms are used. The counselling centre is inevitably identifiable, as there are few such centres in Sylhet, but the cases depicted are not recognizable. The translations from Bengali to English were done by the author and the originals are securely stored.

The researcher responsible for data collection, Laila Ashrafun, is native to Bangladesh and a resident of Sylhet. Her fluency in the local language and dialect and her cultural knowledge were a great advantage in data collection and in building rapport with the interviewees. However, being of middle-class background and a highly educated woman, but not a lawyer made her position ambivalent. Her motives in participating in the mediations and in interviewing the female victims required elucidation, so that she was not mistaken for a representative of the NGO or court personnel. Collecting data in a situation such as mediation, extremely stressful for the participants, turned out to be emotionally taxing also for the researcher.

Case studies are seen as appropriate when the purpose is to 'understand some special people, particular problem or unique situation in great depth' (Patton 1990, 54) and where one can identify cases rich in information – rich in that a great deal can be learned from a few exemplars of the phenomenon in question (Patton 1990, 54). The aim of case studies is the precise description or reconstruction of a case. Cases that are taken as samples should be significant for the research question (Flick 2006). In this article, the case of a young married woman, Lipa, is given extended attention in order to provide an ethnographically layered perspective on the situation of battered women and the mediation processes.

After participating in the mediations, expert interviews (Flick 2006, 165) with six lawyers were conducted. A particular point of interest in these interviews was to establish the interrelationship between the role of expert and private person, as the mediators' personal views on gender roles may affect the mediation process.

The genealogy of domestic violence

The most evident sociocultural factor affecting the practice of mediation is the social legitimacy of the violence perpetrated by husbands and in-laws against young wives. A man has the option of controlling his wife through verbal or physical abuse if she does not succumb to his wishes and this is widely accepted by both men and women (Bates et al. 2004; Sambisa et al. 2011; UNICEF 2011). The accounts of the interviewed women, who lived in slums, reflected this in such comments as: 'I have to bear all the torture of my husband if I want to eat from his income' or 'Everybody will say I am bad if I hit my husband or leave my husband's family because of violence' (Ashrafun 2013). However, the slum women studied did not accept severe domestic violence perpetrated against women. In their everyday life they shared their conjugal problems with each other, although they did not usually interfere in others' domestic disputes unless someone's life was in danger (Ashrafun 2013).

The genealogy of Lipa's experiences of domestic violence and the mediation of her case through the NGO highlight the dynamic of domestic violence among underprivileged women and the process of mediation. Lipa is an illiterate 22-year-old Muslim housewife. She was born in a slum of Sylhet, where her parents had migrated 25 years earlier. Lipa was married when she was 19 to an illiterate three-wheeler driver 10 years her senior. Following marriage Lipa moved to live with her husband's family members in another slum of Sylhet.

Just after one year of marriage, Lipa's husband had kicked her in her abdomen when she had been two months pregnant. She had immediately miscarried and become unconscious, but Lipa's affines had not even called a doctor. When they did not manage to bring Lipa to her senses, Lipa's sister-in-law had finally called her parents on her mobile phone. After having received the bad news they had rushed to Lipa and called a doctor, but Lipa's husband and mother-in-law had requested that they keep silent about the violence. Because of their request they had hid the true nature of the incident from the doctor by lying and saying that Lipa had fallen into a well while bathing. Lipa's elder sister and mother had borne all the expenses for her treatment and had taken her home with them and Lipa had lived with them for two months. Then, her mother-in-law had come to take her back. However, even after this serious incident Lipa's husband had kept on beating and disparaging her. As a result, she had lost her mental balance. Now, Lipa's sister sternly said, 'Lipa was beaten by her husband like a cow. This man should be punished'.

According to Lipa, the fundamental cause of the trouble was her husband's extramarital relationships and inability to provide food for his wife and child. Describing her vulnerability in her affinal home, she said that her mother-in-law was also in the habit of abusing her by using foul language and blaming her for her husband's extramarital relationships, calling her a worthless wife. On a few occasions also her father-in-law had slapped (chor) her because of his son's extramarital relationships. One day he had rebuked his son in front of her so that her husband had become very angry due to the humiliation and had consequently hit her later with a thick stick. Everyone in her marital family had witnessed this torture but no one had stopped her husband. Lipa had felt ashamed (lojja) and in pain (kosto).

In the mediation, the husband denied having had any extramarital relationships. He explained that Lipa was insolent and quarrelled (jhagra) with him so he often lost his temper. Lipa's husband claimed that he was without fault (dosh) and blamed Lipa for all the problems. He tried to rationalize his use of violence by stating that because he was a man he could not tolerate any questioning and arguments, so there were quarrels between them, and he hit Lipa as any normal man with an insolent wife would justifiably act.

Husbands' extramarital relationships and their polygyny, dowry demands, and drug and alcohol addiction play a significant role in the picture of domestic violence among slum dwellers in Sylhet (Ashrafun 2013). The marital problems relating to a husband's extramarital affairs or use of sex workers' services, like in Lipa's case, are common. Among the 20 cases examined in detail here, there were eight cases of husbands with extramarital affairs and numerous cases that included pressure on the young wife's parents to provide more dowry and repeated dowry violence towards the young wife.

Economic and sociocultural constraints

Underprivileged young women's vulnerability is commonly related to household economic insecurity: food insecurity, shelter insecurity, job insecurity, or physical insecurity. In poor households, material constraints include poverty or material deprivation caused by loss of agricultural land and employment due to the changing socio-economic structure or natural calamities such as river erosion, drought, flood, or tornado, migration from rural to urban areas, or death or serious illness of the primary earning member in the family.

The gendered income distribution among labourers was one of the contextual factors that reduced women's options in cases of domestic violence. Because women earn less than men, it was very difficult for Lipa's mother to take her daughter and granddaughter to live permanently with her. Lipa was now in a vulnerable situation. She had been living for the last six months with her elderly mother. As Lipa's mother works as a domestic helper, it is not economically possible for her to look after her daughter and granddaughter for a long time and Lipa's mental and physical condition was too weak for her to engage in full-time employment. Apart from the economic hardship, the social condemnation of a woman living without a male guardian makes such a choice undesirable for women. The women themselves consider a house without a man as vulnerable to break-in, sexual abuse, and loss of honour (Ashrafun 2013).

The practice of early marriage and dowry, women's limited property and inheritance rights, obstacles to women working outside the home due to norms of propriety, and social stigma related to marital separation or divorce all contribute to a structural situation in which uneducated, poor women's options are limited when they encounter domestic violence. Behind these phenomena, we find the patrilineal, patrilocal kinship system in which wife-givers are structurally subservient to wife-takers even in cases of homogamy (Fruzzetti [1982] 1990). According to the fieldwork in two slums in Sylhet, kin relations are governed by the values of patriarchy, although some women have begun to engage more in wage-earning employment and some exceptional women are living without a male guardian (Ashrafun 2013). Although patrilocality and female dependence on males is the prevalent norm, it is not unmitigated.

In mediation, most of the young wives presented themselves as helpless victims. The women emphasized their role as a 'devoted wife' who has virtuously fulfilled her duties. Marital maladjustment is a shame for a woman, as it violates the ideals of dependency of younger women on their husbands and senior affinal kin and of unity between a man and wife. The popular Bengali concept of *ardhangini* refers to the wife as the other half of her husband's body (Rozario 1992; Uusikylä 2000). Whatever the reasons for domestic problems, the woman is often blamed for it, as in the case of Lipa when she was blamed for her husband's extramarital relationships.

Domestic problems always involve in-laws of both sides

In Bangladesh and elsewhere in South Asia, domestic violence tends to involve both kin groups and not only the perpetrator and the victim. However, due to the structural asymmetry in most South Asian kinship systems between wife-givers and wife-takers, the wife's kin is the underdog that tries to succumb to the demands of the husband's kin and to derive prestige from fulfilling this virtuous, subordinate role.

The mediation practiced by the NGO is kin-based. Kin groups come to seek resolution for the marital problems of their members together, as a collective. The problems are understood as being part of wider kin relations. That is why it is considered essential and natural for relatives to be present at the mediations. When domestic violence is involved, it is evident that in some cases the aggressors are also other affines than the husband, for example, the mother-in-law, sister-in-law, or father-in-law. A daughter-in-law is a member of the affinal kin group and thus mediation cannot be described as successful unless the in-laws are prepared to accept her back, even in situations where the young couple lives separately from the husband's parents.

In 13 of the 20 cases, violence was perpetrated against the young wife both by in-laws and the husband. Among these cases, some women suffered from repeated dowry demands, some from the husband's extramarital relations, and others from the domination of mothers-in-law, sisters-in-law, and fathers-in-law. In four cases, it was found that the woman's mother-in-law and sister-in-law were the main instigators of violence against the young wife due to dowry demands, disappointment following the birth of a girl child, the young bride's alleged maladjustment, or general hostility. In seven cases, the husband was the sole perpetrator of domestic violence against his wife.

Lipa's current mediation was the third time her family had come to the counselling centre, this time with an appeal for a divorce, dower, and maintenance costs for the child. Lipa's husband and his father considered Lipa's mother and sister as the true culprits and liars. They accused Lipa's mother for taking her from the affinal home too frequently and not letting her return for several months. If they went to fetch Lipa back they were allegedly insulted (*opoman*). Such accusations are grave in a society where the wife-givers are expected to yield to the wife-takers and to provide courtesy, gifts, and allegiance. Lipa's father-in-law shifted the attention in the mediation away from the alleged violence perpetrated by his son to Lipa's family's disrespectful and shameless behaviour. This turn in the argumentation underlines how kinship structures leave the battered wife's side in a dependent and vulnerable situation morally, socially, and economically.

When Lipa's father-in-law continued to accuse Lipa's natal family members for being the true culprits in the case, Lipa's elder sister became agitated and jumped from her chair, accusing Lipa's father-in-law of being a liar. Lipa's sister brought up the fact that this man's son had kicked her sister in early pregnancy and that Lipa could have died. She regretted that foolishly they had not kept any evidence of this violent act and had then told a lie to the doctor in order to protect Lipa's marriage. She expressed regret that they had acted according to the allegiance expected from wife-givers and had not involved the judiciary system by making a police complaint. If they had done that, they would have now been in a strategically better position to negotiate for Lipa's future.

Avoiding divorce at all cost

Despite some legal reforms, there are still numerous shortcomings in the judicial system in Bangladesh when it comes to women's issues in general and domestic violence in particular. The criminal laws and family laws, taken together, are insufficient and too weak to solve the problems women face, when taking into account women's position and the prevalent structural barriers. The general criminal law has no specific laws on domestic violence against women although some specific acts address issues such as dowry violence, trafficking, and acid throwing. Dealing with 'ordinary' domestic violence against women requires going through criminal law or in case of divorce, through family law. One of the shortcomings is that criminal law does not protect a woman by guaranteeing a right to a 'matrimonial home', nor does it offer her shelter during the proceedings (Ashrafun 2013).

Although some legal avenues are available to battered women, economic and cultural constraints can impede women's access to them. First, the economic resources needed for a legal process are way beyond most socio-economically underprivileged women. Second, women rarely have the social and cultural capital to engage in litigation. And third, the burden on the courts and routine corruption translate to a very slow process and an unpredictable outcome, even if the women should receive monetary and practical support from an organization (cf. in India: Aura 2008; Vatuk 2006). Moreover, women are reluctant to proceed through the criminal court because evidence is hard to gain as witnesses do not corroborate in domestic violence.

In Lipa's two earlier mediations at the counselling centre, nothing had changed despite the in-laws' conciliatory words and agreements. Despite severe violence, Lipa and her natal kin had never reported any incidents to the police. The two advocates acting as mediators still expected that the *shalish* would deter Lipa's husband and in-laws from being abusive, although they knew that two earlier mediations and agreements had not been successful. It was evident that the perpetrators had not considered the mediations a matter of great importance, as they had made no serious effort to change the situation. There was no social pressure or other sanctions to encourage them to abide by the agreement.

In her third *shalish*, Lipa's advocate asked what her own wishes were concerning the case. Lipa replied in a low voice that she did not want to go back to her affinal home. She pointed out that her affines had promised in the counselling centre twice before that they would not beat her or insult her but had not kept their promise. According to Lipa, they were 'hypocrites' (Lipa began to cry). Her sister and mother also burst into tears and became furious. They pointed to Lipa's husband and said, 'One day he will kill Lipa. We will not let her go; we do not want her dead body'.

The male advocate for Lipa's husband pointed out that once again the couple should rearrange their conjugal life because they were not alone: the decision also affected others. He reminded that they had a two-year-old daughter who needed her father. According to the advocate, a woman's life without a husband is very hard, so Lipa should reconsider her wish to divorce and give the in-laws a chance to make good on their word. The advocates are always beside her, he said, and if Lipa's husband makes any more mistakes or acts violently against her they would not hesitate to initiate litigation proceedings.

Also Lipa's own advocate tried to make her understand that she had to think about her daughter's future. She was reminded that as a Muslim woman she could have a divorce whenever she wished, but if she did so she could not return to her conjugal life any longer, even if she felt remorse. The advocate told Lipa that she should not concentrate on other people's (meaning Lipa's mother and sister) emotions, because she was married and she had feelings for her husband and for the whole family of in-laws. The female advocate pointed out that it was a well-known characteristic of Bangladeshi women to want to live in a family (*sangsar*). Lipa's mother might find work for her but how long could she look after both of them, the advocate asked. Her elder sister's husband might not let Lipa live with them, as such an arrangement would be considered shameful in Bangladeshi society. Thus, also her own advocate recommended that Lipa should return to her husband and give him and the parents-in-law one more chance.

In their argumentation, the advocates brought forth a wide spectrum of rationalizations for why Lipa should continue living with her violent husband and abusive in-laws: if she was a virtuous mother and family member who thought of others and not only selfishly of herself, she would yield. Even the national character of women was used as a justification for not leaving her husband. The advocates pointed out the practical, social, and economic difficulties that Lipa and her daughter would face as a divorcee and a divorcee's daughter. The whole gamut of rationalizations placed Lipa's need for personal safety and mental and physical integrity as secondary on the agenda. The need for harmonious kin relations overrode both the need to punish the perpetrators of violence (which was expressed in the mediations by Lipa's mother and sister) and the need for a solution that secured Lipa's future and safety (the concern expressed by Lipa and her mother and sister).

After the two advocates' argumentation, Lipa remained adamant. The advocates continued to persuade her to yield and finally they succeeded. Lipa was under tremendous pressure in the mediation from the two advocates and her in-laws and in practice she was forced to agree to return to her husband. When she agreed, her mother expressed her sadness by exclaiming that this time she feared for her daughter's life. The two advocates harshly intervened and told Lipa's mother to stop forcing Lipa to break her marital vow because, according to them, in Bangladesh society it was very hard for a young woman to live without a husband. Their message was that society did not encourage this practice and the advocates had over the years seen the sufferings of divorced women. Their argumentation builds on the idea that remaining in marriage would secure a woman's health, well-being, and economic security, although this is not the case in practice.

Lipa's case is an example of how a *shalish* does not necessarily end the misery of an abused woman. Lipa relied upon the procedure of *shalish* and did not directly file a litigation case against her husband and in-laws, as she initially did not want to jeopardize her marriage. At the outset she simply wanted the violence to stop and for her husband to give up any extramarital relationships. But as the two previous *shalish* resolutions failed to protect her from frequent abuse she became physically and mentally ill and finally ended up wanting a divorce.

Lipa's was not the only case among the 20 mediations observed in which the battered or maltreated woman was not willing to return to married life, while the advocates, in-laws, or natal relatives pressured her to do so. In another mediation case, Nasrin (a 23-year-old Muslim woman from a slum) clearly expressed her choice for a divorce in a strong voice in the same manner as Lipa above. However, her parents and elder brothers several times requested she recant her decision in order to avoid subjecting herself and the natal family to a very vulnerable situation. In Lipa's case the natal relatives supported her desire to divorce, while in Nasrin's case the natal relatives felt that divorce would jeopardize both her and her children's future.

In some other cases, women had internalized the impossibility of divorce and did not even consider it as an option. In the cases examined here, most of the women and their natal families looked for reconciliation and remedy from domestic violence; they did not want a separation, divorce, or legal punishment for a violent husband or in-laws. Sakina (a 21-year-old Muslim woman from a slum, a mother of two) had been granted a divorce but in a repeated *shalish* she wanted to return to her affinal home for the sake of her two children and due to poverty. Similarly, even after experiencing severe physical torture by her husband and mother-in-law related to continuous dowry demands, Ronjita (a 25-year-old Hindu village woman, a mother of three sons) did not want separation from her husband. She only wanted remedy from dowry violence through the mediation.

The ideals of submissive womanhood, asymmetry between the wife-givers and wife-takers in Bengali kinship, the practical difficulties in arranging a living, and the insecurity of women living without men all lie behind the advocates' and often also natal relatives' and in-laws' attempts to convince wives like Lipa who want a divorce to give up this desire. This pushes women to remain in violent relationships in which the perpetrators rarely face any ramifications. Although divorce is an option according to Islamic laws, in practice it is also simultaneously considered unacceptable in Bangladeshi society for both Hindus and Muslims (Ali 2002; Ameen 2005; Bhuiyan 1991).

Class barrier

In the interviews with the six advocates who engaged in ADR processes, it emerged that they tended to have rather rigid ideas about the interests of the underprivileged women they served in mediations. The organization does not have sufficiently trained staff to counsel victims of domestic violence and only helps them by providing legal advice in the mediations. The advocates have not been sensitivity-trained to reflect on gender asymmetries or to question their own values and conceptions of a woman's position in the family. They were rarely conscious of the need to find out how the female victims themselves saw their situation and instead concentrated on finding a way to convince the woman to return to her affinal family and on creating goodwill in the affines so that they would stop maltreating the young wife. The advocates took for granted the culturally hegemonic view of the wife-takers as the ones entitled to make demands on the young wife and her kin.

Some of the advocates interviewed at the counselling centre expressed the belief that as poor and uneducated, the female victims were incompetent to make rational decisions, whereas they themselves were educated people and as lawyers competent to decide for them. They did not see the class difference and power asymmetry between themselves and their clients as an impediment to interaction and cooperation. However, in the observed *shalishes* class and power difference self-evidently created difficulties for the clients' ability to communicate their own interest in the situation. In a number of the 10 mediations examined the women's voices were suppressed and they were unable to make their point of view heard. The patronizing and at times even chauvinist tone that was evident in some advocates during the interviews and in the mediations may be problematic from the perspective of delivering justice and securing social agency for female victims of the underprivileged class.

Discussion

We have examined above whether semi-official ADR in the form of mediation, *shalish*, used by an NGO in Sylhet, Bangladesh helps to provide relief in cases involving domestic violence against young wives. The example of Lipa's *shalish* and the other cases examined point to numerous problems in the Bangladeshi legal system and also in an ADR as practiced in cases of domestic violence.

In light of the qualitative data examined in this study, ADR does not appear to work to further the interests of the battered, underprivileged women in Sylhet. This is due to three issues: the fundamental problems related to using mediation as a method of dispute resolution in cases of domestic violence; the shortcomings of the judicial system and the limited resources of the third sector in a developing country like Bangladesh; and the gendered and generational imbalance of power in society generally, and its intersections with class, which are manifested in the mediations.

Although we have here examined only one particular NGO and some 20 cases of domestic violence, we have good grounds to believe that as long as ADR is executed along the same lines as in the case of this NGO, the problems inherent in the process will be relatively similar in other parts of Bangladesh and even in other parts of northern South Asia. This is because the widely shared South Asian patrilocal and patrilineal kinship system largely dictates the attitude to domestic disputes and their resolution. The colonial legacy of the judicial system is also shared within the larger region, affecting the ways people seek justice through the formal and informal legal system and the police.

The use of ADR in cases of domestic violence

The use of mediation in domestic violence cases has raised growing objections in Euro-American societies where it is no more recommended in cases of domestic violence. Issues of safety, justice, power imbalances, and the rights of the battered woman when mediation is used instead of court litigation have been brought up as causes of apprehension (Grillo 1991; Salem and Milne 1995; Vestal 2007).

However, such misgivings have not emerged in the use of ADR in family disputes in South Asia, where it is common for the entire families to be involved in the mediation and where reconciliation is strongly recommended over divorce or legal proceedings (Goel 2005). The South Asian hegemonic idea that women are responsible for creating and maintaining harmony in kin relations is manifested in the belief that mediation is a justified method for resolving cases of domestic violence.

The mediators stand for harmony against justice and for a status quo in which young women and wife-givers have fewer rights than men, older women, and wife-takers. The use of ADR should not be continued in cases of long term and severe domestic violence in South Asia.

The shortcomings of the legal process

The shortcomings of the legal system mean that mediation may represent the only real opportunity to reach a solution when poor women face domestic problems. As long as the courts are potentially corrupt, inefficient, and take years to reach a verdict, there is no hope for economically disadvantaged and uneducated battered women to obtain justice through formal procedures. The idea underlying the use of mediation in cases of domestic violence in Bangladesh and elsewhere in South Asia is that such violence is not punishable under criminal law in the same way as other forms of personal violence, but should instead be mediated or reconciled.

Unlike in the traditional mediation practices in Bangladesh and South Asia and beyond, women are heard in the process and can present their own case together with their natal kin. However, unlike in 'traditional' mediation, the mediators are outsiders to the community and consequently do not fully understand the social realities or the options available to their clients. Instead, they represent the educated middle-class national elite, with its own particular ideas of class, gender, and the nation. They stand in a terrain in-between civil society and the judicial system, using their authority to persuade women to abstain from separation and from filing police complaints.

It is characteristic to the ADR process that it lacks enforcement mechanisms and sanctions on the parties involved and instead relies on goodwill and trust (Merry 1982; Nader and Grande 2002). Whenever the perpetrators of violence assure their willingness to give up their destructive behaviour or promise to provide maintenance, there is no sanction involved for those who do not abide by the agreement. The only option for the victim, if the agreement is violated, is to bring the case back to mediation.

Carrying out mediation successfully and in a manner that would benefit the interests of the battered young wives requires resources that are limited in the third sector. The NGO examined here is understaffed and very limited in the services it can offer to its clients. The facilitators are not always well trained to deal with victims and offenders; as lawyers, they are usually knowledgeable about the system of legal regulations, but lack training in human psychology and couple relationships. The importance of securing continued funding from foreign donors leads to a situation in which satisfying donors' administrative targets is a top priority, even if it happens at the expense of clients' needs. This donor-oriented culture drives the legal aid workers primarily to fulfil their allocated quota and ensure the following year's funding (cf. Chowdhury 2011).

The gender and kinship system

ADR's impact depends on the structure of power relations within which it operates (Merry 2002). Accommodating culture in the process has the effect of privileging existing social relations and thus privileging and strengthening existing forms of domination (Cohen 2006). The gendered power imbalance in Bangladesh leads to a situation in which divorced women face grave difficulties in arranging their safety and livelihood (cf. in Bangladesh: Ali 2002; Ameen 2005 and in India: Aura 2008; Vatuk 2006). Due to this imbalance, the advocates rarely consider divorce an option for a battered woman, leaving her to cope in a domestic situation that may be dangerous for her health and survival.

A better functioning ADR process would require that the local community be more actively engaged in the proceedings. For example, a women's local collective could act as a pressure group towards the perpetrators of domestic violence to prevent the violation of agreements (cf. Cohen 2006). Negotiation requires conditions of relatively equal power; otherwise, a negotiation becomes a smoke screen for enforcing power over the powerless (Nader and Grande 2002). The use of ADR should not be an alternative to creating a functioning and just legal system: negotiating without the back-up force of law will not work. The force of a just law has to be available as a last resort (Nader 1979; Nader and Grande 2002). Particularly in cases of domestic violence, there must be a viable option of starting a criminal process at any point of the mediation process.

Acknowledgements

This work was supported by the Academy of Finland [grant number SA138232] and CIMO in Finland. We would like to thank the funding agencies for their generous support. We would also like to thank the two anonymous reviewers for their insightful comments and suggestions.

Notes

1. Other estimates of female victims include, e.g. 70% (Rahman 1999, 2001), 43% (Khan, Rob, and Hossain 2001), 42% (Koenig, Ahmed, and Mozumder 2003) and 67% (Bates et al. 2004).
2. E.g. Ain-O-Shalish Kendro, Bangladesh National Women Lawyers' Association, Women for Women, BLAST.
3. 'Affine' refers to persons related by marriage, i.e. marital relatives.

References

Ali, S. 2002. *Violence Against Women in Bangladesh, 2001*. Dhaka: Bangladesh National Women Lawyers' Association.

Ameen, Nusrat. 2005. *Wife Abuse in Bangladesh – An Unrecognised Offence*. Dhaka: The University Press.

Ashrafun, Laila. 2013. "Seeking a Way out of the Cage: Underprivileged Women and Domestic Violence in Bangladesh." PhD diss., Publications of the Department of Social Research 3, The Helsinki University Press, Helsinki.

Aura, Siru. 2008. "Women and Marital Breakdown in South India. Reconstructing Homes, Bonds and Persons." PhD diss., Research Series in Anthropology, University of Helsinki, Helsinki.

Bates, Lisa M., Sidney Ruth Schuler, Farzana Islam, and Md. Khairul Islam. 2004. "Socioeconomic Factors and Processes Associated with Domestic Violence in Rural Bangladesh." *International Family Planning Perspectives* 30 (4): 190–199.

Bhuiyan, R. 1991. *Aspects of Violence Against Women*. Dhaka: Bangladesh Institute of Democratic Rights.

BIDS (Bangladesh Institute of Development Studies). 2004. *Baseline Survey for Assessing Attitudes and Practices of Male and Female Members and In-Laws Towards Gender Based Violence (Final Report)*. Dhaka: United Nations Population Fund.

Capulong, Eduardo R. C. 2012. "Mediation and the Neocolonial Legal Order: Access to Justice and Self-Determination in the Philippines." *Ohio State Journal on Dispute Resolution* 27 (3): 641–681. http://scholarship.law.umt.edu/faculty_lawreviews/84

Chowdhury, Elora Halim. 2011. *Transnationalism Reversed: Women Organizing against Gendered Violence in Bangladesh*. Albany: State University of New York Press.

Cohen, Amy J. 2006. "Debating the Globalization of US Mediation: Politics, Power, and Practice in Nepal." *Harvard Negotiation Law Review* 11 (1): 295–353.

Flick, U. 2006. *An Introduction to Qualitative Research*. Thousands Oaks, CA: Sage.

Fruzzetti, Lina. [1982] 1990. *The Gift of a Virgin: Women, Marriage, and Ritual in a Bengali Society*. Delhi: Oxford University Press.

Goel, Manju. 2005. "Successful Mediation in Matrimonial Disputes: Approaches, Resources, Strategies & Management." *Delhi Judicial Academy Journal* 4 (i): Delhi Mediation Centre, District Courts of Delhi. http://www.delhimediationcentre.gov.in/articles.htm#introducing

Grillo, Trina. 1991. "The Mediation Alternative: Process Dangers to Women." *Yale Law Journal* 100 (6): 1545–1610.

Jahan, Roushan. 1994. *Hidden Danger: Women and Family Violence in Bangladesh*. Dhaka: Women for Women, A Research and Study Group.

Jahan, Ferdous. 2005. "Gender, Violence, and Power: Retributive Versus Restorative Justice in South Asia." PhD diss., University of Pennsylvania.

Kamal, Mustafa. 2004. "Introducing ADR in Bangladesh." *Delhi Judicial Academy Journal* 3 (ii). Delhi Mediation Centre, District Courts of Delhi. http://www.delhimediationcentre.gov.in/articles.htm#introducing

Khan, M. E., Ubaidur Rob, and S. M. I. Hossain. 2001. "Violence Against Women and Its Impact on Women's Lives – Some Observations from Bangladesh." *Journal of Family Welfare* 46 (2): 12–24.

Koeing, M. S., M. Hossain Ahmed, and A. Mozumder. 2003. "Women's Status and Domestic Violence in Rural Bangladesh: Individual and Community Led Effects." *Demography* 40 (2): 269–288.

Liamputtong, P. 2007. *Researching the Vulnerable*. Thousand Oaks, CA: Sage.

Maru, Vivek. 2010. "Access to Justice and Legal Empowerment: A Review of World Bank Practice." *Hague Journal on the Rule of Law* 2 (2): 259–281.

Merry, Sally Engle. 1982. "The Social Organization of Mediation in Non-Industrial Societies: Implications for Informal Community Justice in America." In *The Politics of Informal Justice Vol. II, Comparative Studies*, edited by Richard L. Abel, 17–44. New York: Academic Press.

Merry, Sally Engle. 2002. "Moving Beyond Ideology: Critique to the Analysis of Practice. 'From the Trenches and Towers' Commentary." *Law & Social Inquiry* 27 (3): 609–611.

Nader, Laura. 1979. "Disputing Without the Force of Law." *Yale Law Journal (Special Issue on Dispute Resolution)* 88 (5): 998–1021.

Nader, Laura, and Elisabetta Grande. 2002. "Current Illusions and Delusions about Conflict Management in Africa and Elsewhere." *Law & Social Inquiry* 27 (3): 573–595.

Patton, M. Q. 1990. *Qualitative Evaluation and Research Methods*. Newbury Park, CA: Sage.

Rahman, Aminur. 1999. "Micro-Credit Initiatives for Equitable and Sustainable Development: Who Pays?" *World Development* 27 (1): 67–82.

Rahman, Aminur. 2001. *Women and Microcredit in Rural Bangladesh – An Anthropological Study of Grameen Bank Lending*. Boulder: Westview.

Rozario, S. 1992. *Women and Social Change in a Bangladeshi Village*. London: Zed Books.

Salem, P., and A. Milne. 1995. "Making Mediation Work in a Domestic Violence Case." *Family Advocate* 17 (1): 34–38.

Sambisa, W., G. Angeles, P. M. Lance, R. T. Naved, and J. Thornton. 2011. "Prevalence and Correlates of Physical Spousal Violence Against Women in Slum and Nonslum Areas of Urban Bangladesh." *Journal of Interpersonal Violence* 26 (13): 2592–2618.

Sharma, M. K. 2004. "Conciliation and Mediation." *Delhi Judicial Academy Journal* 3 (ii). High Court of Delhi. http://www.delhimediationcentre.gov.in/articles.htm#introducing

Steele, F., S. Amin, and R. T. Naved. 1998. "The Impact of a Micro-Credit Programme on Women's Empowerment and Fertility Behavior in Rural Bangladesh." Policy Research Division Working Paper No. 115, Population Council, New York.

UNICEF (United Nations International Children's Emergency Fund). 2011. *A Perspective on Gender Equality in Bangladesh. From Young Girl to Adolescent: What Is Lost in Transition? Analysis Based on Selected Results of the Multiple Indicator Cluster Survey 2009*. Dhaka: UNICEF.

Uusikylä, H. 2000. "The Other Half of My Body: Coming into Being in Rural Bangladesh." PhD diss., Research Reports No. 236, Department of Sociology, University of Helsinki.

Vatuk, Sylvia. 2006. "Domestic Violence and Marital Breakdown in India: A View from the Family Courts." In *Culture, Power, and Agency: Gender in Indian Ethnography*, edited by Lina Fruzzetti and Sirpa Tenhunen, 204–226. Kolkata: Stree.

Vestal, Anita. 2007. *Domestic Violence and Mediation: Concerns and Recommendations*. http://www.mediate.com/articles/vestala3.cfm

Index

Note: Page numbers in *italic* type refer to tables
Page numbers followed by 'n' refer to notes